Europe in Transition: The NYU European Studies Series

Series Editor
Martin A. Schain
New York University
New York, NY, USA

This series explores the core questions facing the new Europe. It is particularly interested in studies that focus on such issues as the process and development of the European Union, shifting political alliances, military arrangements, the impact of immigration on European societies and politics, and the emergence of ethno-nationalism within the boundaries of Europe. The series includes both collected volumes as well as monographs.

More information about this series at
http://www.palgrave.com/gp/series/14976

Veera Mitzner

European Union Research Policy

Contested Origins

Veera Mitzner
Future Earth, The Sustainability Innovation Lab at Colorado
University of Colorado Boulder
Boulder, CO, USA

Europe in Transition: The NYU European Studies Series
ISBN 978-3-030-41394-1   ISBN 978-3-030-41395-8   (eBook)
https://doi.org/10.1007/978-3-030-41395-8

This Palgrave Macmillan imprint is published by the registered company Springer Nature Switzerland AG.
The registered company address is: Gewerbestrasse 11, 6330 Cham, Switzerland

For äiti, a master of "sisu" and my inspiration

SERIES FOREWORD

This is an important book, that tells an essential story of European policy success. The author, Veera Mitzner, argues that the story of the development and financing of a common European research policy is important for two main reasons: first, because this area was not part of the "core competence" of the European Union; second, because, over time, support for research became one of the major concerns of the European Union. Moreover, it is useful to add that support for this policy has been important not only in the hard sciences, but also in the softer social sciences (that are not the focus of this book), where the payoff is much less evident. European funding has not just underwritten research, it has also promoted cooperation among European scholars, universities and research institutions.

This book also focuses on broader themes that go well beyond the question of European integration. The most important of these is probably "...the role of knowledge, not only as a factor of production but also as a means to power and influence in a broader sense." The growing importance of technology in international competition, as opposed to competition for land and natural resources, created favorable context for the development of a common research policy. This favorable context was not inevitable, however, since the EU consists of often competitive nation-states.

Mitzner's archival research at the European and national levels enables her to present the arguments and strategies used by European elites who persistently advocated for a common European research policy. The core of this advocacy was similar to the process in the United States, that support for research was necessary in terms of larger goals. The focus in Europe was on the importance of research for economic expansion, while in the United States advocates focused more on national defense (The National Defense Education Act, for example).

In a larger sense, this study is innovative because it covers the long history of policy development, not just of individual EU states, but also of the EU as a collective entity. It relates the development of policy at the national level to the emerging, and increasingly important institutional framework of Europe. Although this story does not ignore conflict and failure, it is finally a history of remarkable success, from which we can learn a great deal about how the European system operates and evolves. For those of us who have been involved with the development of funding for European research, this study demonstrates what we have experienced. European support and funding is very much focused on scientific cooperation, as well as scientific achievement.

Indeed, Dr. Mitzner's work and career are very much a result of this effort. This book has emerged from a doctoral dissertation written at the European University Institute in Florence. It was written by a talented Finnish scholar, and researched in the archives of numerous EU countries, with the support of scholars throughout Europe.

New York, NY, USA   Martin A. Schain

Acknowledgments

Writing about Europe and the EU while staying relevant is difficult. When I started the research for this book in 2008, the Union looked profoundly different: Great Britain was a solid member state, the populist voices across the continent, while getting louder, were still muted, and not even the euro crisis had shaken the edifice of "the ever closer Union." Studying the EU felt even a little dull and predictable. Integration was proceeding on its unspectacular path as new European legislation was quietly prepared and new member states joined. Whoever looked at the European institutions from a distance saw a functioning and growing bureaucracy, certainly not without flaws and tensions, but something that was steadily and gently molding the European political tissue. Over ten years later, we find ourselves in a different reality. In addition to the euro crisis that suddenly lifted the EU to the news headlines and evoked unprecedented political drifts and passions, there has been a profound challenge with immigration, a war in the EU's own backyard, a rise of euroskeptic and nationalist forces throughout the continent, sudden uncertainty in the transatlantic relations, Brexit, and then of course Covid-19. Moreover, we have ten years less time to reform our economies and societies to avoid the apocalypse of the climate change—a disquieting reality that is gradually dawning on decision-makers and public across the region.

While the EU research policy has emerged from these global and European changes and turbulences with remarkable resilience, it is vital to recognize the evolving context in which the chapters of this book were written. It has been a long journey starting with a PhD dissertation that I defended at the European University Institute (EUI) in Florence, Italy, in

2013. Most of the empirical research used here was conducted during my time at the EUI, which took me to archives, libraries, and fascinating adventures around Europe. This early yet substantial part of this project was guided by outstanding scholars, in particular my excellent thesis supervisor Kiran Klaus Patel, whose commitment, professionalism, insight, and patience I am eternally thankful for. I also received valuable guidance from Bo Stråth, Federico Romero, Martin van Gelderen, and Angela Romano, as well as from Dominique Pestre, who provided important support during my stay as a visiting scholar at the Ecole des Hautes Etudes en Sciences Sociales in Paris. In addition, John Krige and Johan Schot, as members of the Examining Board, provided constructive and helpful advice.

Juhana Aunesluoma, my long-time teacher and mentor at the University of Helsinki, not only offered me a desk and computer at the Network of European Studies during my visits in Helsinki but also graciously invited me to join his team after my graduation. While most of my time during that year and a half went into exploring the end of the Cold War in Northern Europe and teaching European studies to enthused international master's students, there was also some time to write a new chapter on the history of EU research policy that now features in this book. I am thankful for the Academy of Finland for funding my work during all those years in Florence and Helsinki.

When life took me to the United States, my book manuscript traveled with me. I am grateful to Mark Mazower for offering me the unique opportunity to spend a year at Columbia University as a visiting scholar, and Harold James for inviting me to Princeton University as a Departmental Guest. At Princeton and Columbia, I was able to not only discuss my work with exceptional scholars but also use excellent research libraries. Alfred Kordelin Foundation generously offered support during that time.

It was beneficial to my work to look at Europe from a distance. With two subsequent fellowships in Washington, DC, I was lucky to meet great people who were involved in shaping the US science policy and who helped broaden my perspective. In particular, I would like to mention the enlightening conversations with Rebecca L. Keiser from the National Science Foundation, and Albert H. Teich, who not only made a long career in the American Association for the Advancement of Sciences but also canvassed and studied joint European research institutions during the time examined in this book.

The very last words of this book have been written in Boulder, Colorado. In the last few years, my career has taken a turn toward more practical application of research policy through the promotion of sustainability science and knowledge-based decision-making. It has been in response to the recent political and global environmental changes, and due to my desire to support and elevate research communities in a position to generate solutions to pressing sustainability challenges. While that work is only vaguely reflected in this book, I would like to thank Joshua Tewksbury, the Future Earth USA Hub Director and my inspiring manager, for the opportunity to be part of his team and this vital effort.

There have been numerous friends and colleagues that have provided important insights and helped shape this book. Many of them have offered dinners or places to stay during my trips to archives in different parts of Europe or kept me company during long days in the library. I am deeply thankful to Admir, Alanna, Alessandra, Alina, Anna, Anne, Antje, Claire, Corine, Cristiano, Elise, Ere, Frederick, Geert, Isabelle, Jenufa, Jens, Jukka, Katherine, Marcia, Maria G., Maria V., Michi, Minu, Rory, Tuire, and Ulrich for their extraordinary friendship and kindness. I have also relied on the expertise and advice of great archivists and librarians in Europe as well as in the United States while navigating heaps of literature and historical documents. A big thank you to you all.

No book manuscript turns into an actual book without proficient editors, and I have been fortunate to work with a professional crew at Palgrave. Thank you, Rebecca Roberts, Michelle Chen, and John Stegner, for your solid work and patience. I am also grateful to Professor Martin Schein for the opportunity to publish in *Europe in Transition* book series.

And finally, I would like to express sincerest appreciation to my loving husband Scott Mitzner, who patiently took care of our little son during the weekends and late nights that I spent finalizing this book, and who, a few days before the very final submission deadline, asked an innocent but pertinent question: Why do you want to finish it? He had seen my struggle in trying to advance a book project in the margins of busy international full-time jobs and the never-ending responsibilities of a mom of a young child. Well, this is a story that needs to be told and that nobody yet has told, at least not the way I am telling it.

Over the years that the manuscript has been a work in progress, it has gained relevance only as the EU's influence in the European research space has extended and as the urgency of making far-reaching sustainability transitions, informed by science, has become palpable. Furthermore, at a time when the legitimacy of science as well as the desirability of continued European integration is on trial, enhanced understanding of past policies and decisions is vital. If this book can help us gain some clarity and direction, it is worth of all the effort.

CONTENTS

1 Introduction   1

Part I   33

2 Research for Growth: The Ideational Foundations of
Research Policy in Postwar Europe   35

3 A Common Research Policy? Launching the Debate   59

4 Euratom: The Troubled Forerunner of Community
Research Policy   105

Part II   125

5 The Years of Questioning   127

6 COST: Distraction or Progress?   147

7 Contrasting Visions and Continuing Struggle   177

Part III   213

 8 The Return of the Gap   215

 9 Research Policy: A Trailblazer for Institutional Change   229

10 Conclusion and Further Thoughts   257

ABOUT THE AUTHOR

Veera Mitzner is the Future Earth Network Lead at the Sustainability Innovation Lab at Colorado, the University of Colorado Boulder, where she designs and leads global efforts to strengthen sustainability science and knowledge-based decision-making. She holds a PhD in History and Civilization from the European University Institute.

Abbreviations

CCRST	Comité consultatif de la recherche scientifique et technique
CEEP	Centre européen de l'entreprise publique
CERD	Comité européen pour la recherche et développement
CERN	European Organization for Nuclear Research
CES	Conféderation européenne des syndicats
CIRST	Comité interministériel de la recherche scientifique et technique
CoE	Council of Europe
CORDI	Comité de la recherche et du développement industriels
COREPER	Committee of Permanent Representatives
COST	European Cooperation in Science and Technology
CREST	Comité de la recherche scientifique et technique
DG	Directorate général
DGRST	Délégation générale à la recherche scientifique et technique
EARTO	The European Association of Research and Technology Organizations
EC	European Community
ECSC	European Coal and Steel Community
EDC	European Defense Community
EEC	European Economic Community
EFTA	European Free Trade Association
ELDO	European Launcher Development Organisation
EMBL	European Molecular Biology Laboratory
EMBO	European Molecular Biology Organization
EPC	European Political Community
ERA	European Research Area
ERC	European Research Council
ERDA	European Research and Development Agency

ERDC	European Research and Development Committee
ERP	European Recovery Program
ESA	European Space Agency
ESF	European Science Foundation
ESPRIT	European Strategic Programme for Research and Development in Information Technology
ESRO	European Space Research Organization
ETC	European Technological Community
EU	European Union
EUA	European University Association
FEICRO	Federation of European industrial cooperative research organizations
GATT	General Agreement on Tariffs and Trade
GDP	Gross Domestic Product
GNP	Gross National Product
IC	Integrated Circuits
IEA	International Atomic Energy Agency
IT	Information Technology
LERU	League of European Research Universities
NASA	The National Aeronautics and Space Administration
NATO	North Atlantic Treaty Organization
OECD	Organization for Economic Cooperation and Development
OEEC	Organization for European Economic Co-operation
OPEC	The Organization of the Petroleum Exporting Countries
PREST	Politique de la recherche scientifique et technique
R&D	Research and Development
SDI	Strategic Defense Initiative
SGCI	Secrétariat général du comité interministériel pour les questions de coopération économique européenne
UNICE	Union des industries de la Communauté européenne

CHAPTER 1

Introduction

In a Europe threatened by narrow-minded nationalism and looming disintegration, there is a thirst for success stories. The European Union (EU), which is stronger than ever, also looks more vulnerable and exposed than ever, and one might ask where it has, during its over sixty years of existence, done well to serve the self-interested and erratic member states, so different from each other. An area of sustained growth and impact has been the research policy. In 2020, the EU is a major player in supporting and fostering European science. The next EU research funding program, Horizon Europe, will be the largest in the Union's history: almost 100 billion euros will be dedicated to research across the EU area—and beyond. Horizon Europe builds on a continuous expansion of EU research funds and activity and highlights a continent-wide consensus on the benefits of pooling resources and granting the European institutions a significant role in shaping research and science across the national borders. Within just a few decades, the emergence of these activities has radically transformed the European research landscape and changed the way in which research is conducted, funded, discussed, and managed at both national and European levels.

With the EU research arm so strong and generous, it is easy to forget the rockiness of the path that led to the massive budgets and initiatives now bolstering European research. As many other aspects of European integration, for a long time, this success looked unlikely as research remained outside of the core competences of the European Communities

© The Author(s) 2020
V. Mitzner, *European Union Research Policy*, Europe in Transition:
The NYU European Studies Series,
https://doi.org/10.1007/978-3-030-41395-8_1

(EC)[1] and the member states were reluctant to yield sovereignty in such a vital sector. Originally, the three European Communities—which in 1992 were transformed into the European Union—had barely any research policy competence: besides the activities of the European Atomic Energy Community (Euratom) in civilian nuclear technology, the limited activities of the European Coal and Steel Community (ECSC) in the field of coal and steel, and some provisions for agricultural research given to the European Economic Community (EEC), the Treaties establishing these three Communities in 1951 and 1958 remained completely silent on the subject. There was no word about the kind of general research policy the European Union is now so forcefully promoting.

This book explores the contested and perhaps even surprising emergence of European Union research policy. How and why did the Community move to an area that did not belong to its core competences? Where did the idea of a common research policy come from? What were its driving forces? Who were its main advocates? How did its design and objectives evolve over time? What made research one of the major concerns of the current European Union? By seeking answers to these questions, this book contributes to the historiography of European integration and the broader transnational history[2] of post-war Europe. Furthermore, by analyzing the creation of the new forms of governance for European research, it adds to the scholarly discussion on policy-making on science.

In addition to describing the creation of one of the EU's most successful policies while portraying the historical complexity of European integration, this book takes a hard look on the underlying political discourses and the robust mental frames that have enabled certain political paths and that continue to guide political action. Indeed, a central contention here is that a strong and widely shared belief in research as an engine for economic growth constituted the central mobilizing force for EC/EU research policy. Without this specific understanding of science and its societal impact, which emerged from the economic scholarship in the United States and after World War II, and was rapidly popularized in Europe by the Organization of European Economic Co-operation (OEEC) and its successor, the Organization for Economic Co-operation and Development (OECD), it would have been hard if not impossible for the EC to justify political activity in the field of research. During the first two decades after World War II, basically all West European nation-states followed the American example and embarked on a feverish crusade on growth. Consequently, as science was recognized as a source of growth, national institutions were rapidly set up to steer and promote scientific activity. The obsession with growth and competitiveness survived even the brief disillusion with science and expansionary economic policies of the late 1960s and the early 1970s. In fact, by the early 1980s, when economies in Western Europe were staggering, and the IT revolution posed new challenges to the European "knowledge society," the enthusiasm about science as an engine for prosperity appeared stronger than ever. And since the goal of economic prosperity was articulated in the EC founding treaties, the supporters of a common research policy, by framing science as a source of economic growth, were able to move the EC into this new territory. Profiting from the postwar market liberalization, rapid technological change, and prevalent worries about European technological retard vis-à-vis its main commercial competitors, they managed to gain support for initiatives that gradually grew into major policy programs. Soon, the contours of research policy in Europe were permanently transformed.

With its explicit attempt to analyze the process of European integration as a part of a broader European and international history, this book adds to the recent studies challenging the more traditional approach treating the Community like a closed reality standing apart from the rest of the world.[3] As Kiran Klaus Patel has argued, "many of the features of the EC/EU can only be understood if studied in a longer timeframe and against the backdrop of these other settings, rather than in isolation."[4] In research, as in many other fields, the Community was a latecomer that not only borrowed and copied from other international organizations, but also competed and cooperated with them. From the very beginning, the EC/EU research policy was promoted and shaped in a crowded field of national and international activity where its existence and ambitions had to be justified. This book supports the findings of other scholars that show how "European rules and regulations were functionally highly fragmented, as many different organizations dealt with a variety of issues, often for specific sectors and activities."[5] It describes an exciting and hitherto undocumented story of a messily evolving area of European cooperation, where the EC/EU's gradually strengthening position led to the uneasy marginalization of other previously important European venues, and to an emergence of entirely new ways of research policy-making.

Writing this particular success story of European integration could easily lead to a teleological narrative of an ever-closer union, steadily moving toward a predetermined goal.[6] However, by showing that this process was not smooth, one can provide valuable insights into a much-neglected aspect of European integration: setback and failure. Very often, the process of integration diverted from the initially envisioned path and resulted in rather creative formations that not quite complied with the federalist dream, but nevertheless served the purpose of achieving greater European unity. To offer an example, two 1970s efforts by the European Commission to enlarge the Community's research policy activity led to the establishment of new institutions outside the EC structures: in 1971, a total of 19 European countries agreed on the creation of European Cooperation in Science and Technology (COST), a loose intergovernmental framework devoted to easing technological cooperation. Three years later, the Commission lost a struggle for the European Science Foundation (ESF), which came into being as a separate, non-governmental organization, and not an EC institution. Today, both institutions continue their existence in the margins of the EU, whose position as the primary European arena for joint research effort largely goes unrivalled.[7] However, back in the 1970s, they challenged the EC-centric path toward European unity. This book accounts both stories, highlighting the pivotal role of experts and other non-state actors in both enabling and complicating the EC/EU policy-making. While important proposals originated from other institutions and individuals as well, the European Commission—with its exclusive right for making political initiatives at the EC/EU level and its consistent pro-integration ambition—usually took the driver's seat in research policy. Often it was pushing its vision with weak alliances. The lack of support by scientists and their representatives largely explains the Commission's early difficulties in convincing the national governments of a Brussels-centric vision of European research policy. The success of the Commission's plans depended not only on their approval in intergovernmental bargains but also on endorsement among a number of other actors with the power to influence national and European political agendas.

Just as Wolfram Kaiser and Jan-Henrik Meyer have argued, in European integration and the Community politics and policy-making, "various societal actors involved in network-type relations with national governmental and supranational institutional actors were often important for the formation of strategic political alliances, the definition of key political objectives and agendas as well as workable policy compromises."[8] To fully understand the dynamics of European integration, one has to look beyond the national governments and the formal EC/EU institutions and recognize the pivotal role of the representatives of social groups in national societies. In research policy, the role of experts was particularly pronounced. Indeed,

John Peterson and Margaret Sharp, writing about technology policy, have argued it being "hard to imagine that another policy field could be more technocratic or dominated by experts." The same could be said about research policy: as a rule, research policy deals with complicated and very technical issues that often go beyond the knowledge of ordinary politicians and diplomats. Moreover, in the rapidly evolving and future-oriented world of science and technology, national preferences are sometimes hard to define—which leaves the floor open to those who are thought to have the required knowledge and skills in a given subject area.[9] Strong expert participation in the Community decision-making on research, especially in the period covered by this study, can also be explained by the relative novelty and the weak juridical basis of the policy sector.[10] From its very beginning, thus, the Community research policy was outlined in various expert groups and committees, constituted by national administrators, scientists, and industrialists who occupied certain authority in their perspective countries. In a rule, these experts had relatively broad policy mandates to conduct their work, and often, their proposals were adopted with minor if any modifications. This finding supports the recent academic literature emphasizing the importance of seeing the numerous expert groups involved in European-level policy-making more than as technocratic bodies and recognizing their influence on the content of EU policies.[11] We can also observe transnational cross-fertilization of ideas as experts moved between different organizations, such as the OECD, the EC/EU, and national administrations. Furthermore, there was a certain process of institutionalization of expert groups, when cooperation became more formalized and permanent.

Some of these individuals could be seen as constituting an "epistemic community." Peter M. Haas defines the epistemic community as a "network of professionals with recognized expertise and competence in a particular domain and an authoritative claim to policy-relevant knowledge within that domain or issue area."[12] In addition to sharing a set of normative and causal beliefs providing value-based rationale for activity and a cognitive basis for contemplating policy alternatives, these professionals from different disciplines and backgrounds are connected by common inter-subjective criteria for weighting knowledge in their domain of expertise. Moreover, a common policy enterprise, meaning "a set of common practices associated with a set of problems to which their professional competence is directed, presumably out of the conviction that human welfare will be enhanced as a consequence," further features the existence and activity of an "epistemic community."[13] However, the EC research policy expert circles remained narrow and exclusive, and they did not fully represent the broader science and research community that for a long time felt little ownership of the Brussels institution's initiatives. It remained distrustful of the plans emanating from Brussels that were often deemed as bureaucratic and distant to the practical world of scientists, researchers, and innovators. Struggling to convincingly define the "European added value" in research, the proponents of a common research policy encountered substantial competition from existing intergovernmental cooperation and grassroots collaboration, which often was based on established and trusted partnerships and alliances.

Placing important emphasis on ideational and institutional continuity, the book accepts the social constructivist notion that all human activity is determined by specific ideas and assumptions that not only shape interests but also produce the specter of available alternatives and options. Social constructivism focuses on social ontologies including intersubjective meanings, norms, rules, institutions, routinized practices, discourse, communicative action, and collective identity formation.[14] According to John Ruggie, constructivists think that the building blocks of (international) reality "are ideational as well as material; that ideational factors have normative as well as instrumental dimensions; that they express not only individual but also collective intentionality; and that the meaning and significance of ideational factors are not independent of time and place."[15] A core idea is that human actors do not exist independently from their social environment, but that the environment is given meaning through ongoing processes of social construction. These processes are constitutive to our identities and interests that cannot be treated as exogenously given.[16] The constructivist approach is particularly well-suited to the study of long-term political and social changes and the transformative nature of European integration.[17]

A useful concept for understanding the shifting ideational conditions for policy-making is also "policy framings," which Johan Schot and W. Edward Steinmueller define as "interpretations of experience, ordering of present circumstances and imaginations of future potentialities that create the foundations for policy analysis and action and shape expectations concerning potentials and opportunities." Framings evolve over time and extend beyond public policy sphere also influencing the mobilization and activities of non-governmental actors.[18] Another conceptual tool often used in studies of science and innovation policy is "policy paradigm," inspired by the seminal work of Thomas Kuhn on scientific paradigms.[19] According to Peter Hall, a policy paradigm is a "framework of ideas and standards that specifies not only the goals of policy and the kind of instruments that can be used to attain them, but also the very nature of the problems they are meant to be addressing." Hall continues: "Like a *Gestalt*, this framework is embedded in the very terminology through which policymakers communicate about their work, and it is influential precisely because so much of it is taken for granted and unamenable to scrutiny as a whole."[20] In policy studies, paradigms are particularly useful for exploring fundamental shifts in policy, which, according to the evolutionary model of paradigm change, also can occur incrementally.[21] While paradigm might refer to a less heterogeneous and more hierarchical set of ideas than policy frames and discourses,[22] to borrow Grace Skogstad and Vivien A. Schmidt, "[w]hatever the term – policy paradigm, frame, or frame of reference – the significant point is that policy outcomes in a given policy domain will normally be consistent with the prevalent dominant ideas about politically feasible, practical, and desirable policies."[23] This book selectively uses discourses, policy frames, and policy paradigms as heuristic tools to illustrate the shifting ideational context in which the European Community/European Union research policy was initiated, shaped, and advocated.[24]

The book further recognizes that there is no fixed definition of "research policy" or "science policy" and that often these two concepts have been used as synonyms, in particular, if "science" is understood widely, including the broad field of academic activity (similar to German *Wissenschaft*). The content and goals of the EC/EU activity in research have greatly been determined by the dominant understanding of "research policy"—or "research" or "science" in general, as well as by the many meanings of "Europe," the "European Community," or the "European Union." "European research policy" has been and remains a contested reality and subject to continuing disagreement. Here the advice given by François Jacq in his study on French politics of research seems particularly pertinent: instead of assuming the preexistence or the ontological necessity of "research policy" or "science policy," one should rather explore "how conflicting logics about what is *politique de la science* progressively shaped an arena of debates."[25] In other words, the subject is situated in a dynamic discursive framework that is characterized by a multiplicity of ideas, alternatives, and options and defined by fundamental uncertainty and openness, but that all the same enables important continuities to transcend tensions and pressures for change.

The creation of the European Union and the emergence of the EU research policy coincided with an unprecedented expansion of science and technology both in size and in scale. As a result of this growth—unique feature of the postwar period—science and technology also "had a much wider range of societal impacts and implications than they had before the war."[26] Throughout the industrialized world, social, economic, and cultural activities, the natural environment, as well as the knowledge production itself, were radically transformed by a rapid scientific and technological progress. At the same time, the European superiority in science and technology that had largely been unquestioned since the scientific revolution of the seventeenth century, started to erode. New leaders, such as the United States and the Soviet Union, and soon Japan and China, directly challenged or overtook Europe's role as the world's leader in science.[27]

This book challenges the idea of the relative independence of science from the politics of the state. Throughout the history of the modern nation-state, science has been closely related to the practice of power, while scientists[28] have been deeply involved in many core functions of national politics.[29] Also, there have been important changes in the relationship between science and the state. A crucial shift occurred during and after the World War II: from that point on, science was firmly embedded in the national political structures, while the relationship between science and state reached a degree of institutionalization unforeseen in any previous period in history. In the conflict, science had proved its usefulness not only in the development and manufacture of armaments but also in the organization of economic production. Furthermore, the fact that the hostilities found continuation in the antagonisms of the ensuing Cold War, strengthened the position of science at the heart of the national political systems.[30] In the new global conflict, essentially fought by visions, ideas, and images, science became an important symbol of national power and prestige, be it economic, military, or cultural.

The immediate result of this change was not only the revolution in the social role of scientist and the transformation of the independent "savant" to a professional entrepreneur at the service of the state, and in the form in which scientific activities were organized.[31] Also the meaning and the purpose of science changed. For instance, in physics (and later in other fields too), the emphasis of theoretical reflection gradually gave way to increasing instrumentalization; abstract knowledge was replaced by objects and know-how, the search for practical control of phenomena, rather than their understanding, now being the priority. Consequently, the development of science was more and more integrated into its economic and political environment, while "the differences between fundamental research, technical development and the invention of artifacts and techniques became increasingly blurred."[32]

But science as an affair of state was not the product of the World War and the particular international environment of the post-1945 era only. Besides the changes in science itself (the rapid expansion and implementation of scientific knowledge, especially in "hard sciences" such as physics), the evolution of the modern welfare state is crucial too. After World War II, the nation-states in Western Europe radically enlarged their responsibilities and assumed a great deal of new tasks, which involved the creation of new policies and institutions in areas like social security, education, and training. Moreover, maintenance of these institutions being costly, the post-war European nation-state became increasingly dependent on continuous economic growth. This is the necessary context to the emergence of the postwar research policy, intimately linked to the ubiquitous objective of national economic expansion.

The European Community came into being not only as an ambitious peace project to reconcile hostilities on the European continent, but also an economic enterprise attempting to increase growth and wealth in its member countries. So it is not very surprising that when research became identified as a factor in economic success and a central domain of political concern, for many advocates of integration it appeared necessary to also grant the Community some competences in this domain. Although the content of the proposals for expanding the EC's research policy role varied, fundamentally they shared the same rationale that has remained remarkably unchanged over the years: pooling national scientific resources at the European level was intended to contribute to national competitiveness and growth. In addition, the compelling need to adapt to the global structural transformation, increasingly apparent since the mid-1960s, constituted another powerful impulse for the Community activity in the field of research.

This change has been given a variety of different names and interpretations: for instance, historian Charles Maier, writing in 2000, saw the last quarter of the twentieth century as "one of the axial crises of the modern era" undermining the previous territorial order in which bounded space had provided basis for collective political security and economic activity. Marked by worldwide troubles such as the weakening of the hierarchical collective discipline, the reappearance of distributive social conflicts and the dissolution of relative harmonious collaborative industrial relations in the capitalist democracies, the American unwillingness to continue supporting the international monetary regime built up after the World War II, and the emergence of the new economic contenders and militant social movements, the process proved revolutionary both in its depth and scale. Moreover, it was greatly affected by science and technology: "The age of coal and iron, and then, too, of hydrocarbon chemistry, of oil and electricity, of aluminum and copper as well as steel – all still epitomized as late as the 1950s and 1960s by the giant integrated steel mill – was overlaid in fact, and in the public imagination, by the technologies of semiconductors, computers, and data transmission, with a new accepted basis for the creation of private wealth. The concept of hierarchically organized Fordist production based on a national territory was supplanted by the concept, if not always the reality, of globally coordinated networks of information, mobile capital, and migratory labor."[33]

What thus followed was the accentuation of the role of knowledge, not only as a factor of production but also as a means to power and influence in a broader sense. As Susan Strange has put it: "[T]he competition between states is becoming a competition for leadership in the knowledge structure. The competition used to be for territory, when land and natural resources were the major factors in the production of wealth and therefore the acquisition of power for the state... Today, the competition is for a place at the 'leading edge' (as the jargon has it) of advanced technology. This is the means both to military superiority and to economic prosperity, invulnerability and dominance."[34] This was a favorable context for research policy, whose strategic importance rapidly increased toward the end of the century, and became further accentuated after the millennium.

Supporting the findings of historical research that identifies the period around the 1970s as the origin of recent globalization and a formative period of European integration,[35] the contemporary sources used in this book reveal observations of a gradual change, starting around the mid-1960s, intensifying toward the end of the decade, and becoming increasingly apparent during the 1970s. This transformation was of a global scale, but it was experienced in Western Europe as an internal crisis, embodied especially in the end of the post-war economic boom and the consequent and simultaneous social and political turbulence. Crucially, this social construction of external or structural context not only created a need to find responses to altered circumstances; it also helped to define the "politically possible" within the EC polity.[36] During those years, as Claudia Hiepel has put it, the EC "was made stable for globalization by adapting to the global transformation processes." That observation puts the popular interpretation of the 1970s and early 1980s as a troubled period in the history of European integration on its head. A number of policies were developed, often incrementally, which later laid the foundation for the influence of the current day's EU. Rather than stagnation and decline, we can see those years as a "history of empowerment,"[37] which took Europe to the new global era. In research policy, observations of the gradual shift in international order, together with the increasingly dominant view that greater investment and cooperation in scientific research would be a good response to the challenges arising from the new circumstances, effectively determined the political choices.[38]

The following pages abundantly demonstrate how between the mid-1960s and the mid-1980s, a new policy space was born, with durable institutional structures, a devoted group of practitioners and champions, and specific, widely approved ideational frames. By the signature of the Single European Act in 1986, which in the EC marked a watershed moment though the first treaty reform since the 1958 Rome treaties, all that was in place, and while a rapid expansion of activities followed, the fundamentals remained remarkably solid. The 1980s was a period when the Commission's initiatives were given more substance, and a more systematic policy approach was developed to better manage the numerous scattered activities.[39] This book will present compelling evidence showing that all this was possible largely because of the legacy left by earlier efforts. The new initiatives and activities were based on existing institutional structures and past experiences—and what is more, they drew on already formulated ideas and conceptions. One can thus recognize strong path-dependencies that emerged when specific beliefs and practices became consolidated, and when the costs of changing the course simultaneously increased.[40] This shows how exploring the early history of the Community research policy is crucial for understanding the events of the 1980s as well as the EU activity today.

This book is the first in-depth investigation on EU research policy based on extensive archival research including the archives of the European

Commission and the three largest EU member states (France, Germany, and Great Britain). Although the European Commission played a major role in the formulation of initiatives and in consolidating and distributing political ideas, its room for maneuver was ultimately determined by the national governments, which until the 1980s proved reluctant to allow the Community to extend its activities in the field. At the same time, important initiatives originated from the member states. Capturing the dynamics of European integration, in other words, necessitates a careful scrutiny of the formation of the national positions and the constant interplay between national and Community actors, while recognizing that with changing positions and loyalties, the borders of these spheres often became blurred.

The importance of the "systemic" level of decision-making[41] and the central role of experts and other non-traditional actors in the EC/EU research policy make it a particularly challenging field of study. Excluded from high politics and diplomatic spotlights, decision-making and debates on research often took place in realms beyond the established institutional structures, and are therefore difficult to trace. Much was dealt with informally, through personal contacts or within various ad hoc committees and working groups, whose work was only haphazardly recorded. This is why in the study of the EU research policy, the official institutional archives only offer partial answers: the Commission and Council papers contain limited information about the individuals drafting proposals and developing the ideational frames for policy. These sources are particularly silent on the selection of experts, which is a part of the Community decision-making that is not regulated by formal legal rules and where the Commission's General Directorates have significant room for maneuver.[42] A more systematic analysis of the tensions between inclusion and exclusion, than was possible for this book, would be helpful for assessing the scope, nature, and the degree of institutionalization of different networks, studying the power dynamics between them, as well as for understanding the drivers of continuity and change.

The frustration with the Community sources led to research in national and private archives: in Germany, the Bundesarchiv in Koblenz, the Archiv für Christlich-Demokratische Politik in Bonn (mostly the personal papers of Fritz Hellwig), and the archive of the Auswärtiges Amt in Berlin were consulted. In France, the papers of the Secrétariat général du comité interministériel pour les questions de coopération économique européenne (SGCI), the institution responsible for coordinating French European policy, and the Délégation générale à la recherche scientifique et technique (DGRST), both held in the National Archives in Fontainebleau, as well as the documents of the archives of the Ministry for Foreign Affairs in Le Courneuve in Paris and the National Archives in Paris, offered insights to the French views of the Community research policy. Interesting were also the British Foreign Office files, kept in the National Archives in London. These documents revealed often very sharp-eyed observations of the outsider and would-be EC member and helped to depict the role that science and technology played in the first Community enlargement. Given that major parts of this book were initially written in Italy, one might be surprised by the total absence of references to documents from Italian archives. This "omission," however, is not due to a disinterest of the author in the Italian policy of European integration or the common view of Italy as a minor player in the process.[43] Italy was actually often a major advocate of the EC research policy, and therefore a closer exploration of the Italian government's interests and political strategies would have been important. The difficult access to the public archives in Italy, however, resulted in a concentration on countries where the work on original documents is easier. The project would certainly have also profited from research in other national archives, and admittedly, focusing on the three big countries constitutes a certain bias. But with a few exceptions, in the period covered by this study, France, Germany, and the United Kingdom were not only the Community members setting the pace for European integration, but they also were the powerhouses of European science. Chapter 8, which was written later than the other empirical chapters relaying on archive sources, stands out as an exception for the multi-archival approach. While it is mostly based on documents from the Commission and the Council archives, it seeks to include the member state's perspectives whenever possible.

This book has three parts: the first part examines the evolution and proliferation of the postwar concept of "research policy," closely related to the overarching objective of economic growth, and the way in which it was connected to the project of European integration. The focus is on the emergence of the idea of a common research policy rather simultaneously in various political circles in the mid-1960s, and the gradual increase of support for proposals to widen the Community's policy competences in the field. While the first chapter traces the roots of a particular conception of the relations between science and the state, strongly driven by economic considerations, and stresses the central role of the OECD in the proliferation of this very much American idea in Western Europe, the second chapter shows how these models became adopted by the European Commission, and how some French public officials envisaged a common research policy on a European scale. The first Council meeting bringing together the ministers responsible for research from all Community member states in 1967 demonstrates the adhesion on the part of the EC member governments not only to the instrumental and economic conception of research and research policy but also in principle to the prospect of broadening the EC's activities in the field. The creation of the European Commission's General Directorate research, also in 1967, on the other hand, set the necessary institutional foundation for future overtures. These two chapters present the necessary discursive framework for all later debates and decisions on the Community activities in research: they show how the pursuit of growth, the conviction of the economic benefits of science, and, last but not least, the fears of European lag in sectors of high technology constituted crucial imperatives for Community action. The third chapter illustrates how the problems in the existing Community research activities in nuclear energy influenced the plans for a more comprehensive common approach to research. Euratom, whose difficulties exacerbated at the very moment when the first proposals for a general research policy were launched, served as a useful though highly problematic basis for the proposed new activities. These findings support a broader thesis of the book, which is the importance of institutional continuity in European integration.

Part II shows how the plans launched in the latter part of the 1960s were complicated by the continuing disagreement on British EC membership. Moreover, besides the political tensions within the six initial EC member states (France, Germany, Italy, Belgium, the Netherlands and Luxembourg), the changed social, economic, and discursive conditions came to determine the initiatives for a common research policy. Not only the ubiquitous objective of accumulating economic growth and the widespread trust in the ability of science to contribute to that goal, but also the desirability of making large public investments in scientific activity, seemed to be under attack. Even though that assault on science proved less pervasive than some contemporary accounts might lead us to believe, the promoters of the new Community policy were nevertheless forced to reconsider their earlier objectives and strategies. Important reforms were also made in national research systems, imposed partly by economic austerity, and partly by the changing debate of the role of science in society, shaped by the new environmental consciousness and calls for a greater social accountability of science.

All this had implications for the Community, whose success on the issue depended on its ability to adapt to altering political and ideational frameworks. The first chapter of the second part is again intended as an introduction to the ideational context framing the EC debates on research, while the following two chapters account the difficulty of the proponents of an EC research policy to realize their visions in the constant tension between intergovernmental and Community-centric solutions. Chapter 5 examines how the political struggle over the British EC membership transformed the Community's plans of a common research policy into a rather unexpected arrangement: COST saw the daylight as a loose intergovernmental enterprise outside the EC's structures. The story of COST shows how in the EC, different policy issues easily became interlinked, and consequently, even powerful initiatives could be overrun by disputes in other sectors. At the same time, rather than failure, COST should be seen as an excellent showcase for the great flexibility and innovativeness of European integration: as a unique creature, born in very specific political circumstances, it added to the great variety of European cooperation. The next chapter addresses the EC's involvement in the creation of the European Science Foundation, an independent and non-governmental European institution devoted to fundamental research, and another disappointment for the Commission. While proving that the EC had become a European science policy actor to be reckoned with, the struggle over the Community's status in the foundation demonstrated that for its political initiatives to succeed, the EC needed support not only from the national governments but also from those whom the political decisions would fundamentally concern: researchers and their representatives and supporters. If COST was hijacked by national governments, the ESF became a project of national research councils and academies distrustful of the EC's bureaucracy. These "setbacks" notwithstanding, a strong impetus for a more Community-centered vision of research policy prevailed. In addition to the Commissioners Altiero Spinelli and Ralf Dahrendorf, the member governments also, at least in principle, continued to adhere to the idea of increasing and diversifying the EC's activities in research. A visible demonstration of this basic support were the specific formulations included in the final communiqués of the summits of the Hague in December 1969 and Paris in March 1972, and the four resolutions on research adopted by the Council in January 1974. Moreover, the discussions within the EC remained remarkably similar to those of the previous years: as before, the ultimate argument for a new Community policy was its likelihood to contribute to greater economic growth and competitiveness.

A somewhat more nuanced approach to growth and the role of science in society did not take the emphasis away from the continuing pursuit of increasing economic prosperity through research.

The third section of the book discusses the new boost for EC research policy in the early 1980s, a period when the imperatives of the global economy appeared more compelling than ever. It underlines the discursive, ideational, and institutional continuity, and elaborates the book's main argument that EU research policy emerged as a consequence of a gradual consolidation and approval of powerful ideas formulated in the 1960s and 1970s. Ultimately, research policy became a part of the EC's new agenda that was sanctioned by the Single European Act, the first treaty reform since 1958. Research policy and the Single Market belonged to the same political package that toward the end of the decade took the European project into a new level. Chapter 7 analyzes the discursive framework of the early 1980s, again showing the critical interaction between the Community and a wider set of international actors. In the first part of the 1980s, the Community was operating in transformed economic, ideational, and political conditions. The new problems of European economies, the accelerating technological change spearheaded by the IT revolution, and the supposed lead of Europe's two main economic competitors, the United States and Japan, made the old worries about a "technology gap" topical again. In this context, the framing of research policy narrowed down to even more exclusively focus on competitiveness, while developing innovation policies and national systems of innovation became a major political concern. The last chapter explores the creation of the EC's new political agenda for research, which ultimately proved successful. Toward the late 1970s, the Commission grew frustrated with the slow progress with its initiatives and changed its strategy: the launch of the first Framework Program in 1984 as well as the first major Community technology programs (such as ESPRIT) was made in a closer consultation with the industry as well as the research and innovation community. Crucially, the Brussels institution could now rely on a sturdy institutional setting. By the 1980s, an "epistemic community" of national and independent actors and the Commission officials had emerged through regular meetings in various Community committees and working groups. This community shared a robust set of assumptions and beliefs that now could be used for drafting new initiatives. With a political context more favorable for deepening European integration and a strong political discourse advocating for innovation policies, a policy window opened for advancing European level activity in research.

The conclusion chapter not only summarizes the main results of the research but also connects the discussions between the 1960s and 1980s to later political developments. It holds that the development of the Community activity in research has been strongly conditioned not only by the overall framework of European integration and the commitment of national governments and academic communities, but also—and this is crucial—by the broader global technological, economic, and political context as well as the evolving ideational circumstances. This reveals a substantial character of European integration: by relying on powerful political framings and discourses, and the continuity provided by its institutions, the Community/Union was able to move into areas that were not sanctioned by the treaties. This process was incremental and difficult and took place in the shadow of major intergovernmental bargains. As a result, the EC/EU emerged as a central, although contested, agent able to shape dominant political paradigms in Europe. The conclusion chapter not only summarizes the story of an emerging institutional and political reality, which eventually transformed the European research policy landscape, but also makes an argument that to stay relevant, the EU research policy must be capable of breaking with the past and dramatically expand its mission to embrace the social and environmental challenges of the twenty-first century. In fact, with the apocalyptic threat of climate change, and the urgent need for socio-technological transformation at scale, opportunities and imperatives for European-level activity in research might be greater than ever.

LITERATURE

Given the centrality of research in current EU affairs and the dramatic impact of European integration on scientific activity throughout the region, it is surprising that no comprehensive research-based monograph has satisfactorily examined the rise of this particular EU policy. Despite the modest wave of studies published in the middle of 1990s, no serious academic effort has been made to explore the roots of the EU activities in research. Most conclusive of the existing works and the standard reference in almost all accounts touching on the past of the EU research policy is the study of Luca Guzzetti *A Brief History of European Union Research Policy*.[44] Although it serves as a good starting point, the book truly deserves its name: what Guzzetti provides is not an in-depth analysis of all twists and turns in the creation of the political and institutional structures of EU research, but a rather useful summary of the Community's activities in the domain until the mid-1990s. Moreover, Guzzetti's narrative, which mostly relies on published Community sources, repeats the official tale of the Commission as the major entrepreneur in the inevitably advancing train of integration. While providing a solid overview and a helpful starting point for anyone wishing to dig deeper, it is too cursory and a one-sided presentation to satisfactorily explain why and how the Community became the actor that it is today.[45]

Guzzetti's pioneering book was followed by the study of John Peterson and Margaret Sharp on the European Union technology policy. Although Peterson and Sharp highlight the difference between technology policy (that they call, borrowing an OECD definition, "public policies designed to promote technological innovation," or the "application of science and technology in a new way, with commercial success") and science policy ("which aims more broadly to expand and advance systematic knowledge about the natural world"),[46] noting that the emphasis of their study is on the former, they offer some valuable insights to the Community's early effort in both fields. This is mainly because science policy (or research policy) and technology policy have often been treated in the EC as parallel initiatives; the early Community activity was almost exclusively focused on the development of new technologies. Even when blueprints of a more comprehensive approach were drawn up, this conceptual confusion persisted. But Peterson's and Sharp's book is not a historical study: the stress is on EU policy-making in the 1980s and 1990s—the time the book was written—and the parts dealing with earlier years rely heavily on Guzzetti's work.[47]

Peterson and Sharp are not the only political scientists that in the 1990s touched the topic. *Recherche scientifique et construction européenne–Enjeux et usages nationaux d'une politique communautaire* of Laurence Jourdain sheds light on the French aspect in the creation of a common research policy.[48] Even though Jourdain concentrates on analyzing the political room for maneuver of the Community authorities and the influence of their action on national systems, her study is still useful for a historian of the EU research policy. This is not least because of its explicit interest in individual actors as well as in the use of science and technology as a political and discursive resource.

More recently, there also have been a few PhD theses other than the one this book is based on, dealing with the historical aspects of EU research policy. Written in Italian, practically impossible to obtain and consequently ignored in most related literature, the thesis of Filippo Pigliacelli, completed in 2004, examines the European Community's activities in research policy in the years 1949–1971.[49] Fortunately, the fruits of Arthe van Laer's thesis project are more easily accessible. Although the dissertation is mainly on the history of the EU information and telecommunication policies and not exactly the Community's policy in the field of research,[50] van Laer explores links between these two fields. She has also published narrower studies on research policy, mostly from the perspective of the European Commission.[51] The Commission is the centerpiece also in the few articles and books written by public officials personally involved in research policy activities, such as Michel André,[52] Filippo Ippolito,[53] and Pierre Papon.[54] This focus is often shared by the analyses, produced by some political scientists, on the more recent evolution of the EU's activities in the field.[55] Many of them focus on individual Community programs, such as the ESPRIT[56] and BRITE,[57] or on specific EU research policy initiatives. Yet, a study in contemporary history should never ignore the wealth of knowledge that resides in the works of scholars examining the EU from the perspective of the present or more recent past. A good example is Tim Flink's fresh body of work on the establishment of the European Research Council (ERC) as well as on a broader conceptual development of European research policy during the last two decades.[58] Useful is also the research by Maria Nedeva, Linda Wedlin, and others that analyzes the emergence of a European research space from a broader analytical and thematic perspective.[59]

It is interesting that not even the European Atomic Energy Community, Euratom, the first significant Community venture in the field of research, has evoked considerable interest among historians. Although its creation has been studied from various angels,[60] the following troubled years during which the very existence of the organization was seriously questioned, have almost completely been ignored in historical research.[61] This is lamentable, because the crisis of Euratom from mid-1960s onward and the first proposals of an overall Community research policy not only coincided temporally but also were closely linked. As will be shown in this book, the existing institutional structures of Euratom and the question of their future constituted a strong motivation to consider new Community activities.

How, then, to explain this silence of scholars, especially historians, of European integration? Is it the technical and highly specialized nature of policy-making on research that has discouraged researchers from engaging with the topic? To some extent this may be true. A surprising observation made in the course of this project was also that there is lack of not only historical studies on the EU research policy but also national research policies. Undoubtedly, the most interesting works mapping out the development of science policy in the six initial Community member states have been produced by French scholars, such as François Jacq[62] and Julie Bouchard,[63] who engage in fascinating conceptual and discursive analysis of the politics of science in France. Unfortunately, similar studies barely exist on the other EU countries. Historical research on the German research system has remained limited in number and mostly focused on the development of national institutions for supporting research.[64] On Italy, studies are equally low in number, and they mostly concentrate on specific sectors of research such as space and nuclear technology. Moreover, they are published almost exclusively in Italian, which excludes many potential readers.[65] Reflections on government activity in research in Belgium and in the Netherlands are even scarcer.[66] The history of research policy in Britain, the country not yet member in the EC/EU during most of the period covered in this study, but nevertheless an important player due to its status as a would-be member and its reputation as the technological and scientific leader in postwar Western Europe, has not been much more popular.[67]

The French bias is understandable here. Traditionally, France has been the Western European country with most centralized structures of coordinating and funding research. Consequently, the idea of a particular national research policy has been stronger in France than for instance in Britain, where the state has played a clearly more limited role, or in Western Germany, were the responsibilities in the domain have been divided between the federal government and the *Länder*. Of course, centralization is not the only reason for academic interest in analyzing the development of government activity labeled as "research policy": research policy in the United States has been a fairly decentralized affair with tasks distributed across the federal administration. Yet US research policy has been subject of several interesting accounts.[68] In part this can be explained by the strong links between the military and the US public research effort, and the subsequent politicization of the subject.

There is a bulk of literature on the numerous projects realized by the European countries in the forums other than the Community.[69] On cooperation of sectorial nature, one can mention, for instance, the works on the European Organization for Nuclear Research (CERN)[70] and the European Space Agency (ESA),[71] as well as the studies with a more general approach to joint European activities in aeronautics[72] and space.[73] In addition, some multilateral programs, such as EUREKA, have received scholarly attention.[74] John Krige's works on American involvement in the reconstruction of European research offer fascinating insights into the emergence of new European and transatlantic networks of scientists, politicians, and science administrators after 1945.[75] Less attention has been devoted to research activities by European organizations with a more general mandate, such as the OEEC (since 1961 the OECD), the Council of Europe (CoE), and the North Atlantic Treaty Organization (NATO). This reflects the traditionally scant interest in the history of these organizations in general.[76] Benoît Godin's extensive investigations into the evolution of science policy statistics in the OEEC/OECD stands out as a notable exception.[77]

With recent political attention to science diplomacy, there has been some interesting EU-supported work on European science diplomacy, also from a historical point of view.[78] Our perspectives on European integration, science, and technology have also been considerably widened by the works of Johan Schot, Thomas Misa, and the others engaged in the research network *Tensions of Europe: Technology and the Making of the Twentieth Century Europe*. These studies not only encourage scholars to broaden their temporal perspective to the era before World War II but also show how technology has worked as an engine of "hidden integration" traversing both national borders and traditional institutional confines.[79]

All these works have been helpful in understanding and explaining the emergence and evolution of the EU research policy. A thorough historical inquiry into the early years of this policy is vital not only for increasing our knowledge of the complex process of European integration and the changing relationship between politics and science. Understanding the longevity and popularity of the concepts and mental frames established in the debates of the 1960s, 1970s, and 1980s, and the historical trajectories that shaped the policy agenda, is elementary for a critical assessment of the EU's current activity in the field. A perceptive discussion of continuity and change is not trivial: the Union is in an increasingly strong position to diffuse its political ideas and transform the way research is understood, evaluated, and funded in the region. This book invites to an illuminating journey to the roots of a vital policy sector with a remarkable potential for shaping tomorrow's Europe.

NOTES

1. Throughout this book, the European Communities (EC) refers to the three European Communities (the European Atomic Energy Community, the European Coal and Steel Community, and the European Economic Community), which preceded the foundation of the European Union in 1992.

2. Borrowing the definition from Kiran Klaus Patel and Wolfgang Keiser, this book defines transnational history as a "perspective, interested in the analysis of phenomena that transcend nations and nation states and that span territorial border and boundaries." Kiran Klaus Patel and Wolfram Kaiser, "Continuity and Change in European Cooperation during the Twentieth Century," *Contemporary European History*, 27 (2018): 165–182 (174).

3. To name just a few: Kiran Klaus Patel and Wolfram Kaiser, "Multiple connections in European co-operation: international organizations, policy ideas, practices and transfers 1967–92," *European Review of History*, 24 (2017): 337–357; Patel and Kaiser, "Continuity and Change," 165–182; Laurent Warlouzet, *Governing Europe in a Globalizing World. Neoliberalism and its Alternatives following the 1973 Oil Crisis* (London and New York: Routledge, 2017); Kiran Klaus Patel, *Projekt Europa. Eine kritische Gesichchte* (München: C.H. Beck, 2018).

4. Kiran Klaus Patel, "Provincializing the European Communities: Co-operation and Integration in Europe in a Historical Perspective," *Contemporary European History*, 22 (2013): 649–673.

5. Wolfram Kaiser and Johan Schot, *Writing the Rules for Europe. Experts, Cartels, and International Organizations* (Basingstoke and New York: Palgrave Macmillan, 2014), 4.

6. This narrative has been famously challenged by Alan Milward in his book chapter "The lives and teachings of the European saints." Alan Milward, *The European Rescue of the Nation State* (London and New York: Routledge 2000), 318–344. See also Mark Gilbert, "Narrating the Process: Questioning the Progressive Story of European Integration," *Journal of Common Market Studies*, 46 (2008): 641–662. For a more recent analysis on European integration narratives, see Wolfram Kaiser and Richard McMahon (eds.), *Transnational Actors and Stories of European Integration. Clash of Narratives* (London and New York: Routledge, 2019).

7. See the Conclusion chapter.

8. Wolfram Kaiser and Jan-Henrik Meyer, "Beyond Governments and Supranational Institutions: Societal Actors in European Integration," in *Societal Actors in European Integration. Polity-Building and Policy-Making 1958–1992*, eds. Wolfram Kaiser and Jan-Henrik Meyer (Basingstoke and New York: Palgrave, 2013): 1–14 (2). See also Wolfram Kaiser, Brigitte

Leucht, and Michael Gehler (eds.), *Transnational Networks in Regional Integration. Governing Europe 1945–83* (Basingstoke: Palgrave, 2010).

9. John Peterson and Margaret Sharp, *Technology Policy in the European Union* (Basingstoke: Macmillan, 1998) 60–65.

10. Åse Gornitzka and Ulf Sverdrup, "Enlightened Decision Making? The Role of Scientists in EU Governance." *Politique européenne*, 32 (2010), 125–149 (144).

11. Julia Metz, *The European Commission, Expert Groups, and the Policy Process: Demystifying Technocratic Governance* (Houndmills, Basingstoke and New York: Palgrave Macmillan, 2015), 2; Meng-Hsuan Chou and Inga Ulnicane, "Introduction: New Horizons in the Europe of Knowledge," *Journal of Contemporary European Research*, 11/1 (2015): 4–15.

12. Peter M. Haas, "Epistemic Communities and International Policy Coordination," *International Organization*, 46 (1992): 1–35 (3).

13. Haas, "Epistemic Communities," 3.

14. Thomas Christiansen, Knud Erik Jorgensen, and Antje Wiener, "The Social Construction of Europe," *Journal of European Public Policy*, 6 (1999): 528–544 (530).

15. John Gerard Ruggie, "What Makes the World Hang Together? Neo utilitarianism and the Social Constructivist Challenge," *International Organization*, 14 (1998): 855–885 (879).

16. Thomas Risse, "Social Constructivism and European Integration," in *European Integration Theory*, eds. by Antje Wiener and Thomas Diez (Oxford and New York: Oxford University Press, 2009): 145–146; Ben Rosamond, "Discourses of Globalization and the Social Construction of European Identities," *Journal of European Public Policy*, 6 (1999): 652–668 (658).

17. Christiansen, Jorgensen and Wiener, "The Social Construction," 529, 538.

18. Johan Schot and W. Edward Steinmueller, "Three Frames for Innovation Policy: R&D, Systems of Innovation and Transformative Change," *Research Policy* 47 (2018), 1554–1567 (1554).

19. Thomas Kuhn, *The Structure of Scientific Revolutions* (Chicago: University of Chicago Press, 1970).

20. Peter A. Hall, "Policy Paradigms, Social Learning, and the State: The Case of Economic Policymaking in Britain," *Comparative Politics*, 25/3 (1993): 275–296 (279). For a recent application to science policy, see: Gijs Diercks, Henrik Larsen, and Fred Steward, "Transformative Innovation Policy: Addressing Variety in an Emerging Policy Paradigm," *Research Policy* 48 (2019): 880–894 (881).

21. Vivien A. Schmidt, "Ideas and Discourse in Transformational Political Economic Change in Europe," in *Policy Paradigms, Transnationalism, and Domestic Politics*, ed. Grace Skogstad (Toronto: University of Toronto Press, 2011), 36–63.

22. Gijs, Larsen, and Steward, "Transformative Innovation Policy," 881.
23. Grace Skogstad and Vivien A. Schmidt, "Introduction: Policy Paradigms, Transnationalism, and Domestic Politics," in *Policy Paradigms, Transnationalism, and Domestic Politics*, ed. Grace Skogstad (Toronto: University of Toronto Press, 2011), 7.
24. For another insightful analysis of the role of ideas in policy-making, see Daniel Béland, "Ideas, Institutions, and Policy Change," *Journal of European Public Policy*, 16 (2009), 701–718.
25. François Jacq, "The Emergence of French Research Policy: Chance or Necessity?" in *Science and Power: The Historical Foundations of Research Politics in Europe*, ed. Luca Guzzetti (Luxembourg: European Communities, 2000), 153.
26. Andrew Jamison, "Science and Technology in Postwar Europe," in *The Oxford Handbook of Postwar European History*, ed. Dan Stone (Oxford: Oxford University Press, 2014), 631.
27. Ibid., 631–634.
28. Scientists denote here "individuals active in research in public or private scientific and academic institutions in the full range of academic disciplines and fields." Gornitzka and Sverdrup, "Enlightened Decision Making," 126.
29. Susan Strange has regarded the emergence of the modern nation-state as a part of the process of changing knowledge structure, which she defines as a structure that "determines what knowledge is discovered, how it is stored, and who communicates it by what means to whom and on what terms." Susan Strange, *States and Markets* (London and New York: Printer, 1994), 121.
30. Dominique Pestre, "Science, Political Power and the State," In *Science in the Twentieth Century*, eds. John Krige and Dominique Pestre (Paris: CRHST, 1997), 62–67; Jamison, "Science and Technology," 634–635.
31. Dominique Pestre and François Jacq, "Une recomposition de la recherche académique et industrielle en France dans l'après-guerre 1946–1970. Nouvelles pratiques formes d'organisation et conceptions politiques," *Sociologie du travail*, 38 (1996): 264–267.
32. Pestre, "Science, Political Power and the State," 71.
33. Charles S. Maier, "Consigning the Twentieth Century to History: Alternative Narratives for the Modern Era," *The American Historical Review*, 105 (2000), 807–831.
34. Strange, "States and Markets," 132.
35. Claudia Hiepel, "Introduction," in *Europe in a Globalizing World. Global Challenges and European Responses in the "long" 1970s*. Claudia Hiepel ed. (Baden-Baden: Nomos, 2014), 9–23,
36. Rosamond, "Discourses of Globalization," 652–668.

37. Hiepel, "Introduction," 14. See also Anselm Doering-Manteuffel and Lutz Raphael, *Nach dem Boom. Perspektiven auf die Zeitgeschichte seit 1970* (Göttingen: Vandehoeck & Rupprecht, 2008); Éric Bussière, "D'une Europe inachevée à l'affirmation du régionalisme européen dans la mondialisation," In *Europe in the International Arena During the 1970s: Entering a Different World*, ed. Antonio Varsori and Guia Migani (New York and Brussels: P.I.E. Peter Lang, 2011), 43, 48–29.

38. However, from the argument that the common research policy was introduced as a means to cope with the global change, does not result that the process of European integration as a whole or even after the 1980s could simply be explained as a reaction to globalization. For this debate, see e.g. George Ross, "European Integration and Globalisation" In *Globalization and Europe. Theoretical and Empirical Investigations*, ed. Roland Axtmann (London and Washington: Pinter, 1998).

39. Van Laer, Arthe. "Research: Towards a New Common Policy," in *The European Commission 1973–86. History and Memoires of an Institution*, ed. Éric Bussière et al. (Luxembourg: European Commission, 2014); Veera Mitzner, "Research Policy," *The European Commission 1986–2000. History and Memory of an Institution,* ed. Vincent Dujardin et al. (Luxembourg: European Commission, 2019).

40. Paul Pierson, "Increasing Returns, Path Dependence, and the Study of Politics," *American Political Science Review*, 94 (2000): 251–267 (252).

41. According to Peterson and Sharp, at the "systemic" level of decision-making, existing between the "high politics" of ministers and the "sub-systemic" practical, day-to-day management and administration of individual research programs, the Community institutions bargain over policy proposals. Peterson and Sharp, "Technology Policy," 60–62.

42. Gornitzka and Sverdrup, "Enlightened Decision Making," 139.

43. Antonio Varsori, *La Cenerentola d'Europa: l'Italia e l'integrazione europea dal 1946 ad oggi* (Soveria Mannelli, Catanzaro: Rubbettino, 2010), 22.

44. Luca Guzzetti, *A Brief History of European Union Research Policy* (Luxembourg: European Communities, 1995).

45. In 2009 Guzzetti published an article in which he analyzed the later developments of Community research policy. Luca Guzzetti, "The 'European Research Area' idea in the history of Community policy-making," in *European Science and Technology Policy. Towards Integration or Fragmentation?* Eds. *Henri Delanghe, Ugur Muldur, and Luc Doete* (Cheltenham and Northampton: Edward Elgar, 2009).

46. Peterson and Sharp *Technology Policy*, 2.

47. The genesis of the EU technology policy is also discussed in the following works: Roger Williams, *European Technology: the Politics of Collaboration* (London: Croom Helm, 1973); Margaret Sharp and Claire Shearmann,

European Technological Collaboration (London: Routledge & Kegan Paul for the Royal Institute of International Affairs, 1987); Wayne Sandholtz, *High-Tech Europe: the Politics of International Cooperation* (Berkley and Los Angeles: University of California Press, 1992); Thomas C. Lawton, *Technology and the New Diplomacy: The Creation and Control of EC Industrial policy for Semiconductors* (Aldershot: Avebury, 1997); Johan Lembke, *Competition for Technological Leadership. EU Policy for High Technology* (Cheltenham and Northampton MA: Edvar Elgar, 2002).

48. Laurence Jourdain, *Recherche scientifique et construction européenne. Enjeux et usages nationaux d'une politique communautaire* (Paris: L'Harmattan, 1995).

49. Filippo Pigliacelli, "Una comunità europea per la scienza: un 'sogno dei saggi'? Alle origini della politica di ricerca e sviluppo delle Comunità europee (1949–1971)." (PhD diss., l'Università di Pavia, 2004).

50. Arthe van Laer, "Vers une politique industrielle commune. Les actions de la Commission européenne dans les secteurs de l'informatique et des télécommunications (1965–1984)." (PhD diss., Université catholique de Louvain, 2010).

51. Arthe van Laer, "Vers une politique de recherche commune. Du silence du Traité CEE au titre de l'Acte Unique," in *Les trajectoires de l'innovation technologique et la construction européenne = Trends in technological innovation and the European construction*, eds. Cristophe Bouneau, David Burigana, and Antonio Varsori (Brussels and New York: P.I.E. Peter Lang, 2010); Éric Bussière and Arthe van Laer, "Research and Technology or the 'Six National Guardians' for 'the Commission, the Eternal Minor,'" in *The European Commission 1958–72. History and Memories of an Institution*, eds. Michel Dumoulin et al. (Luxemburg: European Commission, 2007).

52. Michel André, "L'espace européen de la recherche: histoire d'une idée," *Journal of European Integration History*, 12 (2006): 131–151.

53. Felice Ippolito, *Un progetto incompiuto. La ricerca comune Europea* 1958–88 (Bari: Edizioni Dedalo, 1988).

54. Pierre Papon, *L'Europe de la science et de la technologie* (Grenoble: Presses Universitaires de Grenoble, 2001); Pierre Papon, "L'Europe de la recherche: une réponse aux défis de l'avenir," *Journal of European Integration History*, 12 (2006): 5–10.

55. See e.g.: Luis Sanz Memémdez and Susana Borràs, "Explaining Changes and Continuity in EU Technology Policy: The Politics of Ideas," in *The Dynamics of European Science and Technology Policies*, eds. Simon Dresner and Nigel Gilbert (Aldershot and Burlington: Ashgate, 2001); Jakob Edler, Stefan Kuhlmann, and Maria Behrens, eds. *Changing Governance of Research and Technology Policy – The European Research Area* (Cheltenham, UK and Northampton, MA: Edvard Elgar, 2003).

56. Wayne Sandholtz, "ESPRIT and the Politics of International Collective Action," *Journal of Common Market Studies,* 30 (1992): 129–151.
57. Jakob Edler, *Institutionalisierung europäischer Politik. Die Genese des Forschungsprogramms BRITE als reflexiver sozialer Prozeß* (Baden-Baden: Nomos, 2000).
58. Tim Flink, *Die Entstehung des Europäischen Forschungsrates: Marktimperative – Geostrategie – Frontier Research* (Weiserswist: Velbrück Wissenschaft, 2016); Tim Flink, and David Kaldewey, "The new production of legitimacy: STI policy discourses beyond the contract metaphor," *Research Policy* 47 (2018): 14–22.
59. Linda Wedlin and Maria Nedeva, eds. *Towards European Science. Dynamics and Policy of an Evolving European Research Space* (Cheltenham and Northampton: Edvard Elgar Publishing, 2015).
60. An illuminative, although slightly outdated, overview of the state of art in Euratom literature is provided in Maurice Vaïsse, "La coopération nucléaire en Europe (1955–1958). Etat de l'historiographie," in *L'énergie nucléaire en Europe. Des origines à Euratom* eds. Michel Dumoulin, Pierre Guillen, and Maurice Vaïsse (Louvain-la-Neuve: Euroclio, 1991). See also Jaroslav George Polach, *Euratom: A Study in European Integration* (Ann Arbor: UMI, 1962); Jaroslav George Polach, *Euratom: Its Background, Issues and Economic Implications* (Dobbs Ferry: Oceana, 1964); Peter Weilemann, *Die Anfänge der Europäischen Atomgemeinschaft: zur Gründungsgeschichte von Euratom 1955–1957* (Baden-Baden: Nomos, 1983); Michael Eckert, "Kernenergie und Westintegration. Die Zähmung des westdeutschen Nuklearnationalismus," In *Vom Marshallplan zur EWG. Die Einigung der Bundesrepublik Deutschland in die Westliche Welt,* eds. Ludolf Herbst, Werner Bührer, and Hanno Sowade, (München: Oldenburg, 1990); Pierre Guillen, "La France et la négotiation du traité d'Euratom," in *L'énergie nucléaire en Europe. Des origines à Euratom,* eds. Michel Dumoulin, Pierre Guillen, and Maurice Vaïsse (Louvain-la-Neuve: Euroclio, 1991); Michel Dumoulin, "The Joint Research Centre (JRC)," in *History of European Scientific and Technological* Cooperation, eds. John Krige and Luca Guzzetti (Luxemburg: European Communities, 1997); Mervyn O'Driscoll, "Missing the Nuclear Boat? British Policy and French Military Nuclear Ambitions during the Euratom foundations Negotiations 1955–1956," *Diplomacy and Statecraft,* 19 (1998): 130–151; Ginevra Andreini, "Euratom: An instrument to Achieve a Nuclear Deterrent? French Nuclear Independence and European Integration during the Mollet Government (1956)." *Journal of European Integration History,* 6 (2000): 109–128; Barbara Curli, "L'esperienza dell'EURATOM e l'Italia. Storiografia e prospettive di ricerca," in *L'Italia nella costruzione europea. Un bilancio storico (1957–2007)* eds. Piero Craveri and Antonio Varsiori

(Milano: Franco Angeli, 2009); Mauro Elli, "A politically-tinted rationality: Britain vs. Euratom, 1955–63." *Journal of European Integration History*, 12 (2006): 105–124. On the role of the United States in the creation of Euratom: Jonathan E. Helmreich, "The United States and the Formation of Euratom." *Diplomatic History*, 15 (1991): 387–410; Gunnar Skogmar, *The United States and the Nuclear Dimension of European Integration* (Basingstoke: Palgrave Macmillan, 2004); John Krige, "The Peaceful Atom as Political Weapon: Euratom as an Instrument of U.S. Foreign Policy in the 1950s." *Historical Studies in the Natural Sciences*, 23 (2008): 5–44.

61. The most comprehensive studies on Euratom's development in the 1960s are those of Laurence Hubert and Lawrence Scheinmann: Laurence Hubert, "La politique nucléaire de la Communauté européenne (1956–1968). Une tentative de définition, à travers les archives de la Commission européenne." *Journal of European Integration History*, 6 (2000): 129–153; Lawrence Scheinman, "Euratom: Nuclear Integration in Europe," *International Conciliation*, 563 (1967): 5–66. Euratom's troubles are also touched in Henry R. Nau, "The Practice of Interdependence in the Research and Development Sector: Fast Reactor Cooperation in Western Europe," *International Organization*, 26 (1972): 499–526. More recently, Euratom's legacy in the Community's later policy-making and the "technical orientation of the EU" has been discussed by Andrew Barry and William Walters. Andrew Barry and William Walters, "From EURATOM to 'Complex Systems': Technology and European Government." *Alternatives: Global, Local, Political*, 28 (2003): 305–329.

62. Jacq, "The Emergence of French Research Policy"; François Jacq, "Aux sources de la politique de la science: mythe ou réalités?" *Revue pour l'histoire du CNRS*, 6 (2002); Pestre and Jacq, "Une recomposition," 263–277.

63. Julie Bouchard, *Commet le retard vient aux Français: analyse d'une rhétorique de la planification de la recherche 1940–1970* (Villeneuve d'Ascq: Presses de Universitaires du Septentrion, 2008). For studies with a more institutional emphasis, see, for example, Alain Chatriot and Vincent Duclert, "Fonder une politique de recherche: les débuts de la DGRST," in *L'Etat à l'épreuve des sciences sociales. La fonction recherche dans les administrations sous la Vème République*, ed. Philippe Bézès et al. (Paris: La Découverte, 2005); Jean-François Picard, *La République des savants. La recherche française et le C.N.R.S.* (Paris: Flammarion, 1990); Denis Guthleben, *Histoire du CNRS de 1939 à nos jours: une ambition nationale pour la science* (Paris: CNRS, 2009). In addition, *La Revue pour l'histoire du CNRS* (1999–2010) offers an extensive selection of academic articles shedding light on various aspects of French research policy: http://histoire-cnrs.revues.org/ (accessed December 1, 2019).

64. Peter Weingart and Niels C. Taubert, eds. *Das Wissenschaftsministerium. Ein halbes Jahrhundert Forschungs- und Bildungspolitik in Deutschland* (Weilerswist: Velbrück Wissenschaft, 2006); Helmuth Trischler, *Forschung für den Markt: Geschichte der Fraunhofer-Gesellschaft* (München: Beck, 1999). The following works provide more general approaches: Helmuth Trischler, "Das bundesdeutsche Innovationssystem in den 'langen 70er Jahren': Antworten auf die 'amerikanische Herausforderung,'" in *Innovationskulturen und Fortschrittserwartungen im geteilten* Deutschland, eds. Johannes Abele, Gerhard Barkleit, and Thomas Hänseroth (Köln: Böhlau, 2001); Helmuth Trischler, "Die 'amerikanische Herausforderung' in den 'langen' siebziger Jahren: Konzeptionelle Überlegungen," in *Antworten auf die amerikanische Herausforderung. Forschung in der Bundesrepublik und der DDR in den 'langen' siebziger Jahren,* eds. Gerhard Ritter, Helmuth Trischler, and Margit Szöllösi-Janze (Frankfurt and New York: Campus, 1999).

65. A general view is offered in Claudio Pogliano, "Images and Practices of Science in Post-War Italy," in *Science and Power: The Historical Foundations of Research Politics in Europe*, ed. Luca Guzzetti (Luxembourg: European Communities, 2000). See also Claudio Pogliano, "Le culture scientifiche e technologiche," in *Storia dell'Italia repubblicana. La trasformazione dell'Italia: sviluppo, e squilibri*. Vol. II/2 (Torino: Einaudi, 1995).

For a more institutional approach, see Lorenza Sebesta and Luca Guzzetti, "Gli aspetti internazionali dell'attività del CNR nel secondo dopoguerra," in *Per una storia del Consiglio Nazionale delle Ricerche*, eds. Raffaella Simili and Giovanni Paoloni (Roma and Bari: Laterza, 2001); Mauro Capocci, "Politiche e istituzioni della scienza: dalla ricostruzione alla crisi," in Francesco Cassata and Claudio Pogliano, eds. *Storia d'Italia 26: Scienze e cultura dell'Italia unita* (Torino: Einaudi, 2011), 267–296. For sectorial cooperation, see Michelangelo De Maria and Lucia Orlando, *Italy in Space. In Search of a Stragegy 1957–1975* (Paris: Beauchesne, 2008); Barbara Curli, *Il progetto nucleare italiano:1952–1964 /* Barbara Curli, *Conversazioni con Felice Ippolito* (Soveria Mennelli (Catanzaro): Rubbettino, 2000).

66. Johan Schot, Arie Rip, and Harry Lintsen, eds., *Technology and the Making of the Netherlands: The Age of Contested Modernization, 1890–1970* (Walburg Pers (Cambridge Mass.): MIT Press, 2010). On Belgium: Kenneth Bertrams, *Universités & Entreprises. Milieux académiques et industriels en Belgique 1880–1970* (Brussels: Le Cri edition, 2006).

67. David Edgerton, *Science, Technology and the British Industrial 'Decline' 1879–1970* (Cambridge: Cambridge University Press, 1996); David Edgerton, "'The Linear Model' Did Not Exist: Reflections on the History and Historiography of Science and Research in Industry in the Twentieth

Century," in *The Science–Industry Nexus: History, Policy, Implications*, eds. Karl Grandin and Nina Wormbs (New York: Watson, 2004); David Edgerton, "Science and the Nation: Towards New Histories of Twentieth Century Britain," *Historical Research*, 78 (2005): 96–112.

68. See e.g.: Bruce L. R. Smith, *American Science Policy since World War II* (Washington D.C.: The Brookings Institution, 1990); Stuart W. Leslie, *The Cold War and American Science: The Military-Industrial-Academic Complex at MIT and Stanford* (New York: Columbia University Press, 1993); Joseph Manzione, "Amusing and Amazing and Practical and Military: The Legacy of Scientific Internationalism in American Foreign Policy, 1945–1963," *Diplomatic History*, 24 (2000): 47–49.

69. Some general accounts on this subject: John Krige, "The Politics of European Scientific Collaboration," in *Companion to Science in the Twentieth Century*, eds. John Krige and Dominique Pestre (New York: Routledge, 2003); John Krige and Luca Guzzetti, eds. *History of European Scientific and Technological Cooperation* (Luxemburg: European Communities, 1997); Cristophe Bouneau, David Burigana, and Antonio Varsori, eds., *Les trajectoires de l'innovation technologique et la construction européenne = Trends in Technological Innovation and the European Construction* (Brussels and New York: P.I.E. Peter Lang, 2010).

70. John Krige, ed., *History of CERN, Vol. I and II and III* (Amsterdam: North Holland 1987, 1990, 1995); Dominique Pestre and John Krige, "Some Thoughts on the Early History of CERN," in *Big Science. The Growth of Large-Scale Research*, eds. Peter Galison and Bruce Hevly (Stanford: Stanford University Press, 1992).

71. John Krige and Arthuro Russo, *Europe in Space 1960–1973. From ESRO and ELDO to ESA* (Noordwijk: ESA, 1994); John Krige, *A History of the European Space Agency, 1958–1987* (Noordwijk: ESA, 2000). See also: Oral History of Europe in Space, a project run by the European Space Agency, with the collaboration of the European University Institute (Historical Archives of the European Union). https://archives.eui.eu/en/oral_history/#ESA (Accessed December 2, 2019).

72. David Burigana, "L'Europe, s'envole-t-elle? Le lancement d'Airbus et le sabordage d'une coopération aéronautique 'communautaire' (1965–1978)," *Journal of European Integration History*, 13 (2007): 91–109; David Burigana and Pascal Deloge, "Introduction. Les coopérations aéronautiques en Europe dans les années 1950–1980: une opportunité pour relire l'histoire de la construction européenne?," *Histoire, Économie et Société*, 4 (2010): 3–18; David W. Thornton, *Airbus Industrie: The Politics of an International Industrial Collaboration* (New York: St. Martin's Press, 1995).

73. Lorenza Sebesta, "Choosing its Own Way: European Cooperation in Space. Europe as a Third Way between Science's Universalism and US Hegemony?" *Journal of European Integration History,* 12 (2006): 27–55; Filippo Pigliacelli, "Italy, ESRO and ELDO," in De Maria and Orlando, *Italy in Space*; Lorenza Sebasta, *Alleati competitivi. Origini e sviluppo della cooperzione spatiale tra Europa e Stati Uniti 1957–1973* (Rome and Bari: Laterza, 2002).

74. Georges Saunier, "Eurêka: un projet industriel pour l'Europe, une réponse à un défi stratégique." *Journal of European Integration History,* 12 (2006): 57–74; John Peterson, *High Technology and the Competition State: An Analysis of the EUREKA Initiative* (London: Routledge & Kegan Paul, 1993); Philippe Braillard and Alain Demant, *Eureka et l'Europe technologique* (Brussels: Bruylant, 1991); Veera Mitzner, "Almost in Europe? How Finland's Embarrassing Entry into Eureka Captured Policy Change," *Contemporary European History* 25/3 (2016): 481–504; Also Sandholtz, *High-Tech Europe,* includes an analysis of the Eureka initiative.

75. John Krige, *American Hegemony and the Postwar Reconstruction of Science in Europe* (Cambridge: MIT Press, 2006). See also John Krige, "NATO and the Strengthening of Western Science in the Post-Sputnik Era." *Minerva,* (38) 2000, 81–108. On the US role in the postwar Europe, see also Giuliana Gemelli, "Western Alliance and Scientific Diplomacy in the Early 1960s: the Rise and Failure of the Project to Create a European M.I.T," in *The American Century in Europe,* eds. Laurence R. Moore and Maurizio Vaudagna (London, Ithaca, N.Y.: Cornell University Press, 2003); Giuliana Gemelli, ed. *The Ford Foundation and Europe, 1950's–1970's: Cross-fertilization of Learning in Social Science and Management* (Brussels: European Interuniversity Press, 1998).

76. Exceptions include: Richard T. Griffiths, ed. *Explorations in OEEC History* (Paris: OECD, 1997); Marie-Thérèse Bitsch, eds. *Jalons pour une histoire du Conseil de l'Europe* (Bern: P.I.E. Peter Lang, 1997); Joséphine Brunner, "Le Conseil de l'Europe à la recherche d'une politique culturelle européenne 1949–1968," in: *Building a European Public Sphere. From the 1950s to the Present,* eds. Robert Frank et al. (Brussels: P.I.E. Peter Lang, 2010); Matthieu Leimgruber and Matthias Schmelzer, eds. *The OECD and the International Political Economy Since 1948* (New York and Basingstoke: Palgrave Macmillan, 2017).

77. E.g.: Benoît Godin, *Measurements and Statistics on Science and Technology: 1930 to the Present* (London: Routledge, 2005); Benoît Godin, "The Making of Statistical Standards: The OECD and the Frascati Manual, 1962–2002. *Project on the History and Sociology of STI Statistics,*" Working Paper No. 39 (2008); Benoît Godin, "Making Science, Technology and Innovation Policy: Conceptual Frameworks as Narratives," *Innovation-*

RICEC, 1/2009; Benoît Godin, *The Making of Science, Technology and Innovation Policy: Conceptual Frameworks as Narratives, 1945–2005* (Montreal: Centre – Urbanisation Culture Société de l'Institut national de la recherche scientifique 2009).

78. Project Insscide website: http://www.insscide.eu/ (accessed December 2, 2019).
79. Thomas Misa and Johan Schot, "Inventing Europe: Technology and Hidden Integration of Europe," *History and Technology*, 21 (2005): 1–19; See also the six volumes in the book series Making Europe: Technology and Transformations, 1859–2000, edited by Johan Schot and Philip Scranton. www.makingeurope.eu; http://www.tensionsofeurope.eu/

PART I

Research for Growth: The Ideational Foundations of Research Policy in Postwar Europe

In the fickle world of the twenty-first century, there is no political ideology, it seems, as persistent and ubiquitous as the pursuit of economic growth. Governments across the globe, regardless of their political standing or type of regime, pursue the same goal of continuous economic expansion. The commitment to growth is also shared by the major international institutions, regional trading blocs, multinational organizations, and even by various civil societies. To be sure, the means of achieving growth vary, as do the ultimate purposes of harnessing higher gross domestic product (GDP) rates. The intermediate objective, however, remains the same: increasing the levels of national economic output as fast as possible. To borrow Stephen J. Purdey's expression, in today's societies, economic growth constitutes a kind of utopian paradigm common to all important actors in modern global polity.[1]

Although the pursuit of growth is a relatively old phenomenon that has dictated human behavior since the Enlightenment, it only became a major political philosophy after World War II.[2] The change occurred with the dramatic expansion of the role of the postwar European nation-state, which, in order to secure its legitimacy, assumed an increasing number of political and social responsibilities. As Alan Milward has argued:

> the post-war state in Western Europe had to be constructed on a broader political consensus and show itself more responsive to the needs of a greater range and number of its citizens if its legitimacy was to be accepted. This

© The Author(s) 2020
V. Mitzner, *European Union Research Policy*, Europe in Transition:
The NYU European Studies Series,
https://doi.org/10.1007/978-3-030-41395-8_2

came to mean attempting a much greater range of tasks. It was in the attempt to reassert itself as the basic unit of political organisation that the parliamentary democratic state came to be a force for higher rates of economic growth.[3]

Consequently, as a crucial characteristic of the legitimating ideology of growth, national income became the new basis of national self-consciousness. Milward writes: "Its growth came to occupy in the national collective psyche of Western Europe the place formerly occupied by the growth of national territory."[4] This obsession with growth naturally also included feverish attempts to discover the possible sources of economic expansion—one of which was soon thought to be scientific research. In the years following World War II, research became increasingly regarded as a fundamental element of production and productivity, while the belief that investing in science and technology would more or less directly result in greater growth started to gain ground. A concrete outcome of these considerations was the emergence of "science policy" as a specific policy domain, closely affiliated to economic policy.

The first part of this book discusses the gradual evolution of the idea of scientific research as an element of production and growth, and the advent of "science policy" defined by the anticipated economic benefits of science. It advances the argument that the economic and instrumental understanding of science embedded in science policy was primarily a phenomenon of the latter half of the twentieth century, and that its transmission in Europe took place in the specific conditions of the US supremacy and the Cold War rivalry that manifested itself also as a contest for growth. It is further proposed that as a result of the powerful lobbying of some international organizations, especially the OECD, this particular view of science became widely endorsed not only among national political and business elites, but also in the European Commission, where it formed the essential ideological basis to the objective of a common research policy.

Stephen Purdey writes that the origins of the growth paradigm[5] reach back about 300 years in time to the beginning of the modern era, commonly associated with the emergence of the idea of progress. The rapid accumulation of knowledge and its application to the material world ushered in a new linear worldview, a core component of which was an expectation of gradual and continuous improvement of the human condition. Although there is no simple way to explain this profound change in public thinking, it has often been attributed to a number of historical factors such as the Italian Renaissance, envisioning the possibility of continuous increase of knowledge and skill; the pursuit of rational and scientific enquiry of the European Enlightenment; the rise of capitalism and the Protestant ethic spurring the process of capital accumulation; the Industrial Revolution enhancing material productivity based on new machines and sources of energy; and last but not least, the emergence of the modern state system oriented toward augmentation of national wealth and power. Moreover, the increasing abstraction of economic activity, further encouraged by the transition of classical to neoclassical economics, constituted a favorable ideational framework for the pursuit of growth. The monetization of the market relations and the commodification of the factors of production, together with the increasingly abstract and mathematical way economy was studied and discussed, created an environment where the gross national product (GNP) could be envisioned to have boundless potential for expansion, without any upper limit.[6] In Robert Collins' words:

> What made the postwar pursuit of growth distinctively modern was the availability of new state powers and means of macroeconomic management dedicated to achieving growth that was more exuberant, more continuous and constant, more aggregately quantifiable, and also more precisely measured than ever before.[7]

After World War II, two particular factors further incited interest in growth in Western Europe[8]: the first was new materialism encouraged by the unforeseen expansion in the manufacture, exchange, and ownership of objects, especially in 1945–1960. No period of similar length in European history has witnessed such growth in purchasing power. As an immediate consequence of this increasing wealth—and increasing expectation of wealth—political success in parliamentary elections became determined by the ability of the governments to meet the voter's expanded material needs and demands. As Alan Milward has put it: "Maintaining the political consensus increasingly came to depend on satisfying the aspirations to this increasing ease of life, usually defined in the 1950s in a rather narrowly materialistic way. The intellectual fashions of the time reflected a similar materialism, at times even an obsession with the world of objects."[9] In the more and more prosperous postwar Europe, thus, sustaining growth became an obligation for any national government wishing to stay in power.

The Cold War constituted the other novel factor strengthening the ideology of growth. In the contest between the East and the West, growth became essential in the governments' attempts to support national security through social stability and wealth. Continuous economic success was necessary to maintain the high military spending deemed vital for guaranteeing national security: "More than ever before," Moses Abramovitz writes, "nations viewed their security and power as resting on economic base. To ensure their independence and safety, they concluded they must grow; if ahead, stay ahead; if behind, catch up." But growth also had another, more direct significance: in the Cold War, growth itself was an area of rivalry. In the contest between liberalism and communism, growth and productivity became the competitive measure between the two alternative economic systems. According to Abramowitz, "[t]he rivalry between the USSR and the United States made each country anxious to prove that its system was capable of producing ever higher material conditions and was therefore worthy of emulation, friendship or even alliance."[10] Consumption also became key in the Western European struggle against domestic communist parties, which, in the years following World War II, enjoyed wide support, particularly in Italy and France.

In the years following World War II, growth was thus etched into the heart of European politics, in both domestic and foreign affairs. "What was really remarkable," historian Michael Postan noted in the late 1960s "was that economic growth was so powerfully propelled by public sentiments and policies. In all European countries economic growth became a universal creed and a common expectation to which governments were expected to conform."[11]

"Growthmanship" had American roots. Robert Collins has demonstrated how, after World War II, a succession of growth regimes, emphasizing growth both as an end itself and as a vehicle for achieving a number of ideological goals, was created by American policy-makers and intellectuals.[12] The war had resolved the ambivalence between the economics of scarcity and the economics of expansion, which had characterized the Depression era. In that moment "[t]he goals of balance and recovery gave way to the pursuit of all-out production and full employment." This change was greatly aided by the war conditions; the defense orders from Europe and the need to arm US military forces added new pressure to production and spending. After 1949, "growthmanship" evolved into a full-fledged policy doctrine widely supported in the US administration. Following the "loss" of China to the communists and the Soviet development of the atomic bomb, growth was more decisively incorporated into the nation's Cold War strategy. President Kennedy embraced the goal of faster growth, with an explicit target of a 5 percent rise in GNP per annum, from the outset of his 1960 presidential campaign.[13]

The idea that scientific research would enhance economic expansion, was authoritatively formulated by economists working on the mysteries of growth. An early pioneer was Joseph Schumpeter, who already in the 1910s emphasized the necessity of technological advance for making economic profit and even distinguished between "invention" (the advance of knowledge useful in production) and "innovation" (the exploitation of such knowledge, the introduction of new products or new methods in commercial activity). For Schumpeter, innovation was an economic activity and a crucial element in the economics of growth.[14] In the mid-1950s, the Schumpeterian ideas received support from American economist Robert Solow, who discovered that, contrary to the prevailing wisdom, only a small percentage of new growth derived from the greater output of labor and increased use of capital. The unaccounted remainder, the "Solow residual," was created by other factors, such as more functional production processes or technological sophistication of the machinery. This discovery prompted economists to study the role of innovative ideas in production and growth.[15] It is also not of lesser significance that Solow happened to be President Kennedy's economic adviser and the author of influential OECD documents setting the parameters for the organization's future policy.[16]

In academic literature, coupling scientific research with economic growth was initially presented as 'the linear model' of innovation. "The linear model," writes David Edgerton, "is usually taken to be something like the following: 'basic' or 'fundamental', 'pure' or 'undirected', scientific *research* is the *main* source of technical innovation; the process of innovation is a sequential one, by which discoveries arising in such research are developed in a *sequence* through applied research, development and so on, to production. Overall, the innovation produced is the main source of economic growth." However, Edgerton also argues that "[t]he lack of clarity, the lack of consensus, or indeed debate over the details of the model is itself indicative that we are not dealing with a worked-out model which anyone ever believed in."[17]

Edgerton might be correct in questioning the existence of a universally accepted concept of a linear model. Yet, the idea of the model became a solid part of postwar American research policy, effectively formulated in the legendary report *Science, the Endless Frontier*.[18] The document was drafted in 1945 by Vannevar Bush, the director of the wartime Office of Scientific Research and Development, and prepared at the request of President Franklin D. Roosevelt to set out a vision of the organization of national research effort after the war. For Bush, there was an inherent tension between the goals of basic research that he pictured as an activity "performed without thought of practical ends," and applied research oriented toward use. However, he also saw an interaction between these two categories, believing that basic research would prove a powerful dynamo for technological progress, when applied research and development converted its discoveries into technological innovations. Bush not only succeeded in establishing a dominant paradigm for understanding science and its relation to technology in the post-1945 world, but also strengthened the ideational premises of the "linear model" with his one-dimensional image of basic research leading to applied research and development.[19]

The power of the idea of basic science producing innovations in the postwar American science policy would never have been as formidable had it not been coupled with the politics of growth. As Bruce L. R. Smith has put it, growth, as much as a reality as a promise, "provided the glue for the system. People supported the consensus [on science policy] in part because all shared the benefits of expansion." Basic science was thus supported not so much because of its ability to increase general knowledge and the understanding of the nature, but rather due to the connection Bush and others had recognized between scientific inquiry and practical applications eventually producing economic return. Smith concludes: "Growth was both a condition and a part of the doctrine, and the two reinforced one another."[20]

Quite logically, increasing productivity and growth became a major objective of the American effort in European reconstruction. The productivity movement, launched by the Marshall Plan, was soon amplified by the Organization for the European Economic Co-operation (OEEC, from 1961 OECD), set up in 1948 to distribute the European Recovery Program (ERP) funds and to coordinate national plans for reconstruction.[21] With the support of progressive US businessmen and liberal government officials, American efforts concentrated on building a massive system of assistance programs intended to foster industrial reform in Western Europe. This policy aimed at supporting the independent recovery of Europe through greater economic integration, industrial development, business reform, and the expansion of consumer markets. An essential inspiration came from the powerful theories of modernization of that period, conveying ideas about linear evolution and convergence toward higher stages of growth.[22] According to Michael E. Latham, "[g]rounded in Enlightenment thinking, nineteenth-century conceptions of national identity, and a sense of America's overwhelming postwar affluence and apparent historical success, modernization theory promised both a framework for objective social analysis and a powerful vehicle for social engineering."[23] Although the outbreak of the Korean War in 1950 temporarily shifted the emphasis away from civilian business toward military production,[24] the American activities had lasting impact: in Western Europe, growth and productivity were increasingly embraced as ubiquitous political and economic goals.

As a matter of fact, in Western Europe, the US appeal was strong. "Transatlantic inspiration to European politics of growth," Michael Postan writes, "came not only from what the USA gave or preached but also from what the USA was...American affluence and American levels of consumption–motor cars, domestic gadgets, and all–were held up as rewards to come. In short, America's very presence provided an impulse to European growth and a measure of its achievements."[25] In crucial measures, the secret of the US power in Europe, thus, rested on its insistence on the individual and subjective dimensions of the recovery-modernization process. With the allure of the American consumption culture, the promise of growth was given very concrete material and personal significance.[26] Therefore, however pronounced the distrust of American imports, "Anti-Americanism," notes Tony Judt "was typically confined to cultural elites whose influence made it appear more widespread than it was."[27]

As growth and productivity constituted important goals of American policy in European reconstruction, the US activities in Western Europe also came to include efforts to foster European science. John Krige has shown how after World War II, the United States attempted to use its advances in science and technology to shape European research agendas, institutions, and the allegiances of scientists to correspond to US political and economic interests. All this reflected the increasingly instrumental status of science as an affair of state and a vehicle for growth: "The coupling of science and foreign policy was symptomatic of the new role that science, and the basic science in particular, had in the post-war period, and of its presumed significance to economic growth, industrial strength, and national security."[28] Although direct support to European science, despite the recommendations made by some sectors of the US administration,[29] never became a significant element in Marshall Aid, European science was boosted in many other ways, for instance, by distributing grants to individual scientists through American private foundations.

The most significant American contribution to the postwar European science policy, however, was in the transmission of specific ideas and policy concepts. The OEEC, and especially its successor organization the OECD, proved instrumental in coupling science with productivity and growth. In the 1950s and 1960s, the OEEC/OECD conducted numerous studies on the subject and developed an original accounting framework used for measuring science.[30] The OECD Frascati Manual, first published in 1962, drew heavily on the methodological guidelines of the US National Science Foundation (NSF) and was intended as a statistical answer or policy tool to contribute to the anticipated economic benefits of science. It was soon established as an international standard.[31]

The transition of the OEEC into the OECD in 1961 granted even more emphasis to research. If the work of the OEEC had been devoted to aiding European economic recovery by means of administering the distribution and utilization of the Marshall Aid, the OECD became a broader organization dealing with economic and welfare concerns in more general terms. Consequently, growth assumed an increasingly central position on the organization's agenda. In November 1961, the OECD adopted an explicit target of increasing the combined GNP of the OECD economies by 50 percent in a decade.[32] The organization also engaged a serious study of different factors that were thought to be lying behind economic growth. In this context, a Committee for Scientific Research was established, trying to bring the OECD's work on science in line with the new growth and development objectives.[33]

From now on, the OECD fully began developing the concept of science policy. In 1961, the Wilgress report, named after its author Dana Wilgress, the former Canadian Ambassador to the OEEC and NATO, concluded that: "[t]oo much emphasis cannot be placed on the importance of scientific efforts for the future of the European economy" and that "[t]he first thing should be for each country to draw up a national science policy." In Wilgress' view, this national science policy would essentially concentrate on science as a basis for technological innovation and economic performance.[34]

In 1961 the Secretary-General of the OECD, Thorkil Kristensen, appointed an ad hoc group to advise him on what the place of science should be in the expanded context of the reformed organization. In addition, the group was asked to consider the implications of science for economic growth and also the more general relationship of science and technology to the formulation of public policies in all fields.[35] The subsequent report, "Science and the Policies of Governments" (the so-called Piganiol report, named after its author Pierre Piganiol), was published two years later. It offered a set of recommendations for the member governments to follow in supporting scientific and technical research. Crucially, the document, considered as a starting point of the organization's activities in the area,[36] emphasized the idea that science, together with higher education, should be seen as a productive factor on par with labor and capital in the pursuit of growth.[37] The group recommended the organization of a conference that would bring together the responsible ministers from all the OECD countries. That event took place in Paris on October 3–4, 1963, and marked the first major occasion where science policy was discussed multilaterally at a ministerial level. In the meeting, consensus seemed to emerge on the instrumentalist perception of science as an auxiliary to economic policy, a view that figured also in the final communiqué: "The Ministers recognized the growing importance of science and technology in the economic and social development of the Member countries. They therefore stressed the importance of establishing effective links between science policy and economic policy."[38]

The OECD's work was critical, because in the early 1960, science policy associated with economic considerations was still largely an unknown territory for many Western European governments. The Piganiol report from 1963 noted that "[t]he idea of an explicit national science policy is new...because science has only recently taken on major public dimensions."[39] Accordingly, in the first OECD ministerial meeting on science, several national delegations stressed the novelty of the idea.[40] Alexander King, an influential OECD official at the time, recalled from the event:

> Then there were doubts expressed by some countries of the wisdom of convening such a gathering in the dominant economic environment of the OECD. The Dutch Minister of Education came to Paris for the express purpose of persuading [Secretary-General] Thorkil Kristensen to abandon the project. In a conversation over lunch that day, he denounced the association with the economy as a "prostitution of science," and demanded that it be regarded of educational or cultural policy.[41]

These concerns were addressed within the OECD. In 1965, the OECD ad hoc Advisory Group on Science Policy, attempting to "define and explore the new area of governmental concern that is coming to be known by the shorthand term 'science policy,'" stressed "not only the implications of science for economic growth, but also the more general relations of science and technology to the formulation of public policies in all fields."[42] However, the focus remained on economy, and it did not take long for this new concept of science policy to find its place in Western European politics. When the second OECD ministerial meeting was organized in February 1966, "there was a much greater understanding of the relationship between science and the economy. Even fundamental research was recognized as a long-term investment both for the economy and for injecting vitality into the educational system."[43] While in 1963 only four OECD countries had a minister of science, in 1966 already two-thirds of the organization's member governments were represented by a person holding such a portfolio.[44] Almost overnight, science policy had become one of the core functions of the Western European states.

In part, the European responsiveness to the new idea of science policy stemmed from the worries of the technology gap. The origins of the notion of the gap have often been attributed to the French Délégation générale à la recherche scientifique et technique (DGRST), a governmental body responsible for scientific research. It first appeared in 1964 in an article of Pierre Cognard, published in *Le Progrès scientifique*. Cognard wrote: "Numerous are those who think that…Europe is on the point of making up for its slowness compared to the United States.…[Unfortunately, they are basing themselves] on a somewhat outdated conception of productive wealth, dating back to an age when the classical factors of production were only capital, manpower and primary materials." In Cognard's view, "a new step in industrial revolution was underway which will be marked by a systemic use of scientific progress in industry." In this new context, the American superiority risked "creating a science gap to the benefit of the United States."[45] By creating a similarity between the "dollar gap" of the late 1940s, the "missile gap" of the late 1950s,[46] and the OECD notion of the "productivity gap,"[47] the "technology gap" became a popular and long-standing image and strong incentive to increase European research effort.

Crucially, the notion of the gap was based on three general assumptions: the first was the one propounded by the OECD Frascati manual, namely, that scientific performance could be quantitatively measured, and that comparisons could be made between countries. The second assumption was that sooner or later, there would possibly be economic and technological convergence between individual nations. The third contained the core idea of the postwar research policy, namely, that economic growth and productivity were desirable goals and derivative from scientific research. All these three assumptions won wide acceptance among Western European political and economic elites during the 1960s and became established as enduring paradigms guiding political and industrial activity in the domain of research.

That gap fitted so well with the dominant ideas of science, economy, and international relations, explains, in part at least, its speedy proliferation. While in the minutes of the OECD ministerial meeting of October 1963, Europe's "delay" is mentioned only once,[48] by the time of the third reunion in 1968, the gap was on everybody's lips. Perhaps most media attention to the gap was attracted by *Le défi américain*, the best-selling book of Jean-Jacques Servan-Schreiber, published in November 1967.[49] In his book, Servan-Schreiber, the co-founder of the French weekly *Express* and one of the most influential opinion-makers in France, managed to capture the concerns about the presumed transatlantic technological difference and sell them to a wider public in an attractive and polemic form. His ideas found immediate resonance in a number of the leading Western European newspapers, to the extent that an American diplomat could lament that it was now impossible to avoid the subject in European press.[50]

There was also another reason why the gap soon gained such popularity: it found a fertile ground in the national narratives of "delay" and "decline," especially abundant in the former European colonial powers, where the loss of international influence was most strongly felt. For Great Britain, the idea of the gap matched well with the arguments of a national economic and technological downturn since the 1870s. This discourse, aligned with the old tradition of British "declinism,"[51] was prompted by the loss of the British innovative leadership. Its central arguments were people's lack of enthusiasm for science and technology, a low social status of the scientists and engineers, and the indifference of the government to the needs of technology.[52] Toward the late 1950s, the debate around "decline" accelerated, simultaneously becoming increasingly centered on the idea of Britain lagging behind the rapid rise of living standards elsewhere in Western Europe. The following decade barely challenged the rationale behind declinist arguments. On the contrary, in the run-up to the 1964 parliamentary elections, the theme came to occupy a prominent place in the rhetoric of the major political parties.[53] Moreover, these discussions were supported by the talk of "brain drain" that embodied British worries about what looked like an alarming growth in the emigration of British scientists and engineers to the United States. This concept, which stemmed from an influential Royal Society Report and was commonly associated with the British Minister of Science, Lord Hailsham, remained in the active political vocabulary of the leading Labour politicians even after the election of the Labour government in 1964. In October 1967, a report conducted on the request of the government even called "brain drain" a national threat.[54]

A similar debate on "delay" had a long tradition in France. Julie Bouchard has demonstrated how the discourse of "retard" constituted a determinant element in the formulation of the French politics of research. According to Bouchard, a common observation in the 1950s was that the French lagged behind other nations in relation to scientific manpower and the development of certain scientific disciplines. In the following decade, insufficiencies were also registered in a number of other disciplines. At the same time, the notions of "retard" became more closely linked to the idea of international economic competition.[55]

Common to the French and British discourses is the strong emphasis on cultural arguments. In Britain, the argument commonly put forward to explain poor national performance in innovation and the use of technology was the traditional hostility of the British elite toward technology and technological education.[56] The French "tardophiles," in contrast, depicted an anti-capitalist and conservative France, badly adapted to the rules of dynamic and triumphant capitalism.[57] These images reflect the thinking of American sociologist William F. Ogburn who, by recognizing lags between the exponential growth of inventions and the ability of the society to adapt to technical change, created an influential narrative that was repeated in the major policy documents and debates in the United States in the 1930s.[58]

Almost always, the "decline" or "retard" was perceived in relation to the rising Western power, the United States of America. The first wave of alarmist European books on "Americanization" dates from the turn of the century and expresses anxieties unleashed by the growing American exports of industrial machinery, farm equipment, hardware, and other engineering and producer's goods.[59] After World War II, these sentiments were considerably strengthened, also in countries where the process of decolonization and the loss of great power status were less dramatic than in Great Britain and France. "What American society displayed repeatedly," writes David Ellwood, "was its capacity to create, produce, and distribute desirable visions of progress on an unrivalled, industrial scale. This was a power potential almost fully lacking in the Old World most of the time, and one that brought back to traditional elites their original nineteenth-century worries about industrial and commercial capitalism, but now with a new intensity and a specific place in mind."[60]

The fear of American technological ascendancy was of course not totally unfounded. After World War II, it really seemed as if the scientific and technological balance in the world had irreversibly shifted from the old continent to the other side of the Atlantic. Historian Thomas P. Hughes has described the United States as the first technological nation that was brought into being during the century after 1870 and that was characterized by particular technological enthusiasm. The era of inventing, developing, and organizing large technological systems—production, communication, and military—was driven by a firm assumption of the possibility of human beings to create a world of their own design.[61] After World War I, the American zest for inventors and innovation contrasted dramatically with the questioning of science and technology among European intellectuals, disillusioned with the destructive effects of the war. "In the decades after World War I," Michael Adas writes, "applied science and technology pervaded American life to a degree that greatly exceeded that experienced by any other society throughout history." Very soon, "Americans came to regard invention and technological innovation as endeavors in which they could surpass all other peoples, including the British who had played vital roles in the industrialization of the United States."[62] By 1945, the United States had secured its position not only as the world's most productive economy by virtually any measure but also the undisputed technological leader. In 1963–1964, the Americans spent nearly four times more money on R&D than the Western European countries altogether, and five times more than the six European Community members.[63] In addition, in 1967 there were three times more people working in research-related fields in America than in Europe, while the number of European scientists and engineers (148,000) remained clearly inferior to that of American (436,000).[64]

An essential product of the American technological nation and its World War II effort to realize, within a few years, the research, development, and construction of the two atomic bombs, was the concept of "big science."[65] At times there were even 250,000 people participating in the

Manhattan project—a scheme that required a critical amount of central planning and steering.[66] This model persisted also after the war in the growing number of large-scale, coordinated research projects aimed at serving both public and private interests.[67] In the postwar years, governmental interest in the big science projects was crucially spurred by the specific conditions of the Cold War and the symbolic, political, and economic role science came to play in that new global confrontation. In particular, after the launch by the Soviet Union of the world's first artificial satellite Sputnik in 1957, an event that dramatically exposed the capacity of Soviet science and technology, the Federal Administration made determined efforts to increase US scientific output. The most spectacular example of the government engagement in research is perhaps the Apollo program, which, running from 1961 to 1972, culminated in the first manned Moon landing in 1969.

The postwar rise of American R&D is impressive, but it alone does not explain why in the middle of the 1960s, Europeans so anxiously started to look across the Atlantic. As we have seen, the concerns about American ascendancy were not new, and even the figures pointing out the American advancement in some sectors of R&D had been common knowledge for a while.[68] Moreover, by the mid-1960s, the differences between the United States and Europe were already rapidly diminishing.[69] The debate on the gap coincided with the postwar boom years, a period characterized by unparalleled growth across the Western European countries.[70] In France, for instance, between 1960 and 1973, the average annual economic growth was 5.5 percent, which was actually higher than the growth rates in the United States. Also, research funding during the decade witnessed a remarkable expansion: the share of research of France's GDP went from 0.79 percent in 1958 to 2.23 percent in 1967.[71] The trend was very similar in West Germany, which doubled its R&D spending between 1962 and 1971.[72]

To a large extent, the worries about European delay might—paradoxically—be explained by the increased wealth of Western European states. In the years following World War II, European governments had been forced to concentrate their resources on the most pressing tasks of the ruined continent. Only now, it was possible to sacrifice time and money to less urgent issues. As a matter of fact, the gap emerged rather simultaneously with growth debate. Matthias Schmelzer has noted that in the immediate postwar years, mainly due to other policy concerns, the idea of economic growth was pretty much absent in the political discussions.[73]

Since the pursuit of growth underpinned the gap debate, it is not surprising if there was little serious concern about European retard before the definitive appearance of growth politics. But wealth in Western Europe contributed to panic about the gap also in a less direct way: in the 1960s, the renewed European strength and the consequent willingness of the Europeans to pursue their own interests effectively loosened the transatlantic alliance.[74] This was manifested for instance in the attitudes toward the rapidly increasing US investment in Western Europe: while during the 1950s, the European governments had virtually been competing to attract American capital,[75] in the course of the following decade, American investment was seen in a more negative light. Particularly in France, the worries aired by Jean-Jacques Servan-Schreiber and others about the dominance of US industry found strong resonance.[76] Last but not least, the comparative studies conducted by the OECD on national R&D activities essentially increased the awareness of relative differences in scientific performance between the member states. Ultimately, the gap thus emerged through the calculations of public officials, who lent the necessary prestige and credibility to tables showing an indisputable European "delay."

Regularly coupled with notions of urgency, the gap was a powerful mobilizing image that could be operationalized for various political purposes.[77] More often than not, the debates of a decline or delay were part of a particular political agenda with the objective of increasing the government support for science and technology. David Edgerton writes that in Britain, "scientists and engineers themselves were the key propagators of the argument. Convinced that *national* investments in science and technology are the key sources of *national* economic power, they necessarily explained what they saw as deficiencies in economic power by a lack of science and technology or its misapplication."[78]

But the politics of the gap were not limited to the national context. Right after its inception, the gap became a political imperative in European integration. The basic argument was straightforward: science had expanded exponentially and now required resources that not even the largest European states could provide on their own. In the world of "big science," national efforts remained too fragmented and insufficient, while European firms lacked the critical scope to successfully face global competition. So, for the promoters of integration, the problem of the gap was first and foremost a problem of scale. The American advantage not only stemmed from generous government spending, but also from the sheer size of the country and its powerful companies, which were big enough to undertake large-scale research and enter large markets. As the following chapters will demonstrate, this argument was adopted by a number of politicians, public officials, and other actors promoting a "common research policy" within the framework of the European Community.

In the context of the Cold War, it is interesting how the technology gap debate constructed a fuzzy mental category of (Western) "Europe" opposed to an American "other" and thus departed from the conventional Cold War reasoning that identified the Soviet Union or the "East" as the only serious rival. For Maurice Duverger, writing in *l'Express* in 1964, Communism was no longer a threat: "There is only one immediate danger for Europe, and that is the American civilization."[79] Stephen Graubard noted the same year:

> Europe's continuing concern is with the possibility of becoming too much the political, intellectual and spiritual disciple of the United States. The opposition to Americanization–which is a sort of short-hand for the complex trends toward modernization–…expresses a sense of the necessity which Europeans feel, that they must continue to be something which Americans are not. This is a strange reversal of roles. Europe, in the twentieth century, is seeking the independence from the other which America so prided itself on securing in the eighteenth.[80]

In the 1960s, the allure of the American promise "you can be like us," started to sound more like a menace than a generous promise of a rising superpower. At the same time, the growing concern of European independence anticipated a situation where international relations were defined, not along the Cold War bloc lines, but in terms of global economic and technological rivalry, where scientific knowledge had accentuated importance.

All this aside, the technology gap was not necessarily an anti-American concept. As such, the notion of the gap contained only the idea of *difference* between Europe and the United States—all the rest being a matter of interpretation. Although for most Europeans, this difference was undesirable, it was not clear that Americans would benefit from the gap either. In fact, Americans proved rather sensitive to European worries.[81] Science being a vital asset in the Cold War confrontation, Western European weakness in the domain could prove disastrous to the entire alliance. Therefore, transatlantic technological concurrence was a positive-sum game: the point for Europe was not to reduce American scientific and technological power but rather to catch up, to be at least as good as Americans.

Its appeal notwithstanding, the gap remained a vague and disputed concept. To a lesser extent, this was also the case with "science policy" or "research policy." Ministers and public officials continued to have great difficulty in articulating what policy in the field of research was actually about, and in what sense it would differ from earlier governmental efforts in the domain. In 1963, Pierre Piganiol stated in his report that: "[t]he phrase 'science policy' is used [here] no more frequently than necessary because of the ambiguity of its meaning. ...When used, it is, where possible, with the narrow connotation of a policy specifically for the advancement of science."[82] Science policy came into being as a discursive construction with a nebulous and malleable content.

Yet during the 1960s, an understanding of science policy or research policy as an auxiliary to economic policy started to gain ground in Western Europe. As Benoit Godin has put it: "from its very beginning, science policy, whether implicit or explicit, was constructed through reflections on accounting, economic growth, productivity, and competitiveness."[83] Leaving military concerns aside, it was in the pursuit of economic goals where, especially after the early 1960s, Western European governments mostly found the meaning for their activity in the domain of science.

To sum up, from the early 1960s onward, the debate surrounding the technology gap proved a discursive foundation for the creation of national science policies intended to boost economic growth. The emergence of the overreaching goal of growth as a major political principle of the post-war era, and the linking of science to the objective of expansion, fundamentally transformed the relations between science and the state in Western Europe. This specific framing of science policy constituted the necessary ideational basis for the development of the European Community as a research policy actor. Without that new economic rationale of research policy and the surge of the European fears of a transatlantic lag, it is unlikely that the Community would have started to discuss an initiative that found no juridical basis in the founding treaties.

Notes

1. Stephen J. Purdey, *Economic Growth, the Environment and International Relations, The Growth Paradigm* (London and New York: Routledge, 2010), 3–4.

2. Ibid., 63.
3. Alan Milward, *The European Rescue of the Nation State* (London and New York: Routledge 2000): 27.
4. Ibid., 27.
5. "[T]he growth paradigm is a world order in which the role played by ideas has taken precedence over the role played by either material capabilities or institutions, both of which can be conceived as expressions of, or at least as compatible with, the dominant ideational feature of the era, as originally expressed in the Enlightenment conception of progress." Purdey, *Economic Growth*, 87. For a more recent study on the emergence of the growth paradigm, see: Matthias Schmeltzer, "The growth paradigm: History, hegemony, and the contested making of economic growthmanship," *Ecological Economics*, 118 (2015): 262–271. See also Matthias Schmeltzer, *The Hegemony of Growth. The OECD and the Making of the Economic Growth Paradigm* (Cambridge: Cambridge University Press, 2016).
6. Purdey, *Economic Growth*, 26, 66, 73–75.
7. Robert M. Collins, *More. The Politics of Economic Growth in Postwar America* (Oxford: Oxford University Press, 2000), xi.
8. The first country to make maximizing the growth rate a political goal was actually the Soviet Union. Decisive growth policies were practiced also by the European fascist governments that promoted technological modernization and showed remarkable creativity in designing institutions for the management of the economy for the purpose of growth. Volkmar Lauber, "Ecology Politics and Liberal Democracy," *Government and Opposition*, 13 (1978): 210.
9. Milward, *The European Rescue*, 129.
10. Moses Abramovitz, *Thinking About Growth and Other Essays on Economic Growth and Welfare* (Cambridge: Cambridge University Press, 1989), 11.
11. Quoted in David W. Ellwood, "The Marshal Plan and the Politics of Growth," in *Explorations in OEEC History*, ed. Richard T. Griffiths (Paris: OECD, 1997), 104.
12. Collins, *More*, xi.
13. Ibid., 10–25, 49; David W. Ellwood, *Rebuilding Europe. Western Europe, America, and Postwar Reconstruction* (London and New York: Longman, 1992), 223.
14. Abramovitz, *Thinking About Growth*, 9–10. The basics of Schumpeter's theory of innovation can be found in Joseph A. Schumpeter, *The Theory of Economic Development: An Inquiry Into Profits, Capital, Credit, Interest, and the Business Cycle* (Cambridge M.A.: Harvard University Press, 1934—first published in 1911). On the evolution of Schumpeter's ideas on innovation, see John Hagedoorn, "Innovation and Entrepreneurship: Schumpeter Revisited," *Industrial and Corporate Change*, 5 (1996):

883–896. On Schumpeter's life and thinking in more general, see Wolfgang F. Stolper, *Joseph Alois Schumpeter: the Public Life of a Private Man* (Princeton: Princeton University Press, 1994); Elias G. Carayannis and Christopher Ziemnowicz (eds.), *Rediscovering Schumpeter* (Basingstoke: Palgrave Macmillan, 2007); Thomas K. McCraw, *Prophet of Innovation: Joseph Schumpeter and Creative Destruction* (Cambridge M.A.: Belknap Press, 2010).

15. Purdey, *Economic Growth*, 76.

16. Schmeltzer, "The Growth Paradigm," 268.

17. David Edgerton, "'The Linear Model' Did Not Exist: Reflections on the History and Historiography of Science and Research in Industry in the Twentieth Century," in *The Science–Industry Nexus: History, Policy, Implications*, eds. Karl Grandin and Nina Wormbs (New York: Watson, 2004): 32.

18. Vannevar Bush, *Science, the Endless Frontier*, A report to the President by Vannevar Bush, director of the Office of scientific research and development (Washington D.C., United States Government Printing Office, 1945).

19. Although coining the concept of "basic research," Bush was not the first person making this differentiation that has deep roots in the Western tradition of science and philosophy. Most effectively, the separation between the two categories of research was instrumentalized in the nineteenth-century Germany, where universities devoted to investigation of "pure" science, and *Technische Hochschlen* (technically oriented higher education institutions) and industrial research institutes conducting applied research, constituted an influential system that strongly inspired Americans. Donald E. Stokes, *Pasteur's Quadrant. Basic Science and Technological Innovation* (Washington D.C.: Brookings Institution Press, 1997), 2–4, 36–39.

20. Bruce L. R. Smith, *American Science Policy since World War II* (Washington D.C.: The Brookings Institution, 1990): 38–39.

21. Ellwood, "The Marshal Plan and the Politics of Growth," 99–105.

22. Mark H. Haefele, "Walt Rostow's Stages of Economic Growth: Ideas and Action," in *Staging Growth, Modernization, Development and the Global Cold War*, eds. David C. Engerman et al. (Amherst and Boston: University of Massachusetts Press, 2003), 81–103.

23. Latham, Michael E, "Introduction: Modernization, International History, and the Cold War World," in *Staging Growth, Modernization, Development and the Global Cold War*, eds. David C. Engerman, Nils Gilman, Mark H. Haefele, and Michael E. Latham (Amherst and Boston: University of Massachusetts Press, 2003): 2.

24. Jacqueline McGlade, "From Business Reform Programme to Production Drive, The Transformation of US Technical Assistance to Western Europe," in *The Americanisation of European Business the Marshall Plan and the*

Transfer of US Management Models, ed., Ove Bjarnar and Matthias Kipping (London and New York: Routledge, 1998), 30–31.

25. Quoted in: Ellwood, *Rebuilding Europe,* 226–227.
26. Ibid., 227. On the arrival of American consumer culture, see also Victoria de Grazia, *Irresistible Empire: America's Advance Through Twentieth-Century Europe* (Cambridge, M.A.: Belknap Press, 2005).
27. Tony Judt, *Postwar, A History of Europe since 1945* (London: Vintage Books 2010), 353.
28. John Krige, *American Hegemony and the Postwar Reconstruction of Science in Europe* (Cambridge M.A.: MIT Press, 2006), 3.
29. The inclusion of science in Marshall Aid was advocated, for instance, by the Research and Development Board led by Vannevar Bush, who regarded strong scientific capacity as necessary for sustained economic growth in Europe. Ibid., 16, 30–31.
30. Benoît Godin, "The Value of Science: Changing Conceptions of Scientific Productivity, 1869 to Circa 1970," *Social Science Information,* 48 (2009): 547–586. Of course, the OECD was not alone dealing with these questions. The problem of science and growth was also studied in other international organizations, such as UNESCO, which in 1960 adopted a "Ten-Year-Plan" in exact and natural sciences with the explicit aim of assisting member countries to raise their scientific level and extend the applications of science to their economic growth. Jean-Jacques Salomon: "International Scientific Organisations," in *Ministers Talk About Science,* 57, 64–65. The OECD, however, very soon assumed the role as the leading international institution in this field, while UNESCO's activity gradually diminished. As Peter Tindemans has noted, "UNESCO's impact on the development of the organization and policies for science in Europe has never come near the influence of the OECD and virtually disappeared in the 1970s." Peter Tindemans, "Post-war Research, Education and Innovation Policy-Making in Europe," in *European Science and Technology Policy, Towards Integration or Fragmentation?* Eds. Henri Delanghe, Ugur Muldur, and Luc Doete (Cheltenham and Northampton: Edward Elgar, 2009), 10.
31. Benoît Godin, "The Making of Statistical Standards: The OECD and the Frascati Manual, 1962–2002. *Project on the History and Sociology of STI Statistics,*" Working Paper No. 39 (2008): 3, 5, 15, 51.
32. Emmanuel G. Mesthene, "Introduction," in *Ministers Talk About Science,* ed. Emmanuel G. Mesthene (Paris: OECD, 1965): 27; Schmeltzer, "The Growth Paradigm," 267.
33. King, Alexander, "Science in the OECD," in *Ministers Talk About Science,* ed. Emmanuel G. Mesthene (Paris: OECD, 1965): 19–20.
34. Ibid., 23–24.

35. Mesthene, "Introduction," 27–28.
36. Muriel Le Roux and Girolamo Ramunni, "L'OECD et les politiques scientifiques. Entretien avec Jean-Jacques Salomon," *La Revue pour l'histoire du CNRS*, 3 (2000).
37. "The clear conclusion is that a large measure of economic growth in advanced countries is the result of new knowledge produced by scientific research, and of the material capital that embodies it." OECD, *Science and the Policies of Governments* (Paris: OECD, 1963), 27.
38. Mesthene, *Ministers Talk About Science*, 132.
39. OECD, *Science and the Policies of Governments*.
40. Procès verbal de la Conférence Ministérielle sur la Science, Paris, le 3 et 4 octobre 1963, BAC 3/1978-786/3-4, Historical Archives of the European Union, Florence (hereafter HAEU).
41. Alexander King, "Scientific Concerns in an Economic Environment: Science in OEEC–OECD," *Technology in Society*, 23 (2001): 343.
42. Mesthene, *Ministers Talk About Science*, 27–28.
43. King, "Scientific Concerns," 343.
44. Julie Bouchard, *Commet le retard vient aux Français: analyse d'une rhétorique de la planification de la recherche 1940–1970* (Villeneuve d'Ascq: Presses de Universitaires du Septentrion, 2008), 218; King, "Scientific Concerns," 343.
45. Quoted in Benoît Godin, *Measurements and Statistics on Science and Technology: 1930 to the Present* (London: Routledge, 2005): 224. See also Pierre Gognard, "Recherche scientifique et indépendance," *Le Progrès Scientifique*, 76 (1964): 1–15.
46. Lorenza Sebasta, "Un nuovo strumento politico per gli anni sessanta. Il *technological gap* nelle relazioni euro-americane," *Nuova Civiltà delle Macchine*, 67 (1999): 11–23.
47. Godin, *Measurements and Statistics*, 223.
48. OCDE: Conférence Ministérielle sur la Science. Procès-verbal de la Conférence réunie au Château de la Muette, à Paris le jeudi 3 octobre et le vendredi 4 octobre 1963, BAC 3/1978-786/3-4, HAEU.
49. Jean-Jacques Servan-Schreiber, *Le défi américain* (Paris: Denoël, 1967).
50. John W. Young, "Technological Cooperation in Wilson's Strategy for EEC Treaty," in *Harold Wilson and European Integration: Britain's Second Application to Join the EEC*, ed. Oliver J. Daddow (London: Frank Cass, 2003), 105.
51. For the pervasiveness of the declinist debate, see Richard English and Michael Kenny, "British Decline and the Politics of Declinism," *British Journal of Politics and International Relations*, 1 (1999): 252–266.
52. David Edgerton, *Science, Technology and the British Industrial 'Decline' 1879–1970* (Cambridge: Cambridge University Press, 1996): 1–2.

53. Jim Tomlinson, "Inventing 'Decline': The Falling Behind of the British Economy in the Postwar Years," *The Economic History Review*, 49 (1996): 732.

54. Matthew Godwin, Jane Gregory, and Brian Balmer, "The Anatomy of the Brain Drain Debate, 1950–1970s: Witness Seminar," *Contemporary British History*, 23 (2009): 35–60.

55. Bouchard, *Commet le retard vient aux Français*, 35–36.

56. Edgerton, *Science, Technology*, 2.

57. Bouchard, *Commet le retard vient aux Français*, 52–53.

58. Godin, "The Value of Science," 3–4.

59. As in the later declinist debate often was the case, also these images were partially exaggerated, particularly in the first half of the twentieth century: although in some sectors, especially in steel industry, Americans had gained a leading position by that time, the United States was the world leader neither in science nor in the use of science-based technologies. Richard R. Nelson and Gavin Wright, "The Rise and Fall of American Technological Leadership: The Postwar Era in Historical Perspective," *Journal of Economic Literature*, 30 (1992): 1931–1964.

60. David W. Ellwood, *The Shock of America: Europe and the Challenge of the Century* (Oxford: Oxford University Press, 2012), 2. This is of course not to say that comparisons would not have been made also between European countries. Especially in Britain from the 1950s, continental Europe was increasingly seen as natural point of comparison, a tendency greatly strengthened by the movement toward economic union and the possibility of British participation in this project. Tomlinson, "Inventing 'Decline,'" 743.

61. Thomas P. Hughes, *American Genesis* (New York: Viking, 1989), 1–11.

62. Adas, Michael, "Modernization Theory and the American Revival of the Scientific and Technological Standards of Social Achievement and Human Worth," in *Staging Growth, Modernization, Development and the Global Cold War*, eds. David C. Engerman, Nils Gilman, Mark H. Haefele, and Michael E. Latham (Amherst and Boston: University of Massachusetts Press, 2003): 25, 30, 33.

63. Even without military spending, the difference continued to be important. Jean-Jacques Sorel (Jean-Jacques Salomon), "Le retard technologique de l'Europe," *Esprit*, 365 (1967): 765.

64. Sorel (Salomon), "Le retard technologique de l'Europe," 902.

65. As a concept, however, "big science" was established only in the 1960s. I am thankful to Geert Somsen for pointing this out.

66. On the history of the concept of "big science," see Helmuth Trischler, "Das bundesdeutsche Innovationssystem in den 'langen 70er Jahren': Antworten auf die 'amerikanische Herausforderung,'" in

Innovationskulturen und Fortschrittserwartungen im geteilten Deutschland, eds. Johannes Abele, Gerhard Barkleit, and Thomas Hänseroth (Köln: Böhlau, 2001), 58. See also Lew Kowarski, "Psychology and Structure of Large-Scale Physical Research," *Bulletin of the Atomic Scientists,* 5 (1949): 186–204; Alvin M. Weinberg, *Reflections on Big Science* (Cambridge, M.A.: MIT Press, 1967); Derek J. de Solla Price, *Little Science, Big Science* (New York: Columbia University Press, 1963).

67. Peter Galison, "The Many Faces of Big Science," in *Big Science, The Growth of Large-Scale Research,* eds. Peter Galison and Bruce Hevly (Stanford, C.A.: Stanford University Press, 1992), 1–4.

68. Henry R. Nau, *National Politics and International Technology, Nuclear Reactor Development in Western Europe* (Baltimore and London: The John Hopkins University Press, 1974), 44, 50.

69. Nelson and Wright, "The Rise and Fall of American Technological Leadership," 1931–1964.

70. Mark Mazower, *Dark Continent. Europe's Twentieth Century* (New York: Vintage Books, 1998), 292–298.

71. Bouchard, *Commet le retard vient aux Français,* 135.

72. Helmut Trischler and Hans Weinberger, "Engineering Europe: Big Technologies and Military Systems in the Making of 20th Century Europe," *History and Technology,* 21 (2005): 64.

73. Schmeltzer, "The Growth Paradigm," 266.

74. Geir Lundestad, *The United States and Western Europe since 1945. From "Empire" by Invitation to Transatlantic Drift* (Oxford: Oxford University Press, 2003), 113; Hubert Zimmermann, "Western Europe and the American Challenge: Conflict and Cooperation in Technology and Monetary Policy, 1965–1973," *Journal of European Integration History,* 6 (2000): 85–110.

75. In 1950 the US investment stood at 1.7 billion dollars representing about one-seventh of US investment abroad. By 1970 it has grown to 24.5 billion, which was about one-third of total US investment abroad. Lundestad, *The United States and Western Europe since 1945,* 112.

76. Ibid., 134.

77. Bouchard, *Commet le retard vient aux Français,* 137.

78. Edgerton, *Science, Technology,* 68–69.

79. Quoted in Judt, *Postwar,* 353.

80. Stephen H. Graubard, "A New Europe?" *Daedalus,* 94 (1964): 543–566.

81. Richard R. Nelson, "World Leadership, the Technology Gap and National Science Policy," *Minerva,* 9 (1971): 386–399.

82. OECD, *Science and the Policies of Governments,* 13.

83. Godin, "The Value of Science," 182.

CHAPTER 3

A Common Research Policy? Launching the Debate

One of the main aims of the European Economic Community (EEC) was to increase the wealth and prosperity of its members. This growth objective was already included in the Article 2 of the Rome Treaty, defining the purpose of the EEC: "The Community shall have as its task, by establishing a common market and progressively approximating the economic policies of Member States, to promote throughout the Community a harmonious development of economic activities, a continuous and balanced expansion, an increase in stability, an accelerated raising of the standard of living and closer relations between the States belonging to it."[1] It is thus not very surprising if the supporters of European integration were attentive to the international debates on science policy, mostly emanating from the OECD. If scientific research held out the promise of growth, it would be important for the Community to become active in that field too.

This chapter shows how the instrumental view of science embracing the imperatives of productivity and growth was taken up and gradually consolidated by the advocates of European integration. It also demonstrates how the European Community (EC) started to carve out a place for itself in the increasingly crowded field of European cooperation in science and technology. By the mid-1960s, international research cooperation had become an objective of remarkable political interest in Western Europe. "Barely a quarter of a century ago," wrote Jean-Jacques Salomon, the secretary of the Interim Committee in the OECD Directorate for Scientific Affairs in 1965, "international scientific cooperation and the problems it created

© The Author(s) 2020
V. Mitzner, *European Union Research Policy*, Europe in Transition:
The NYU European Studies Series,
https://doi.org/10.1007/978-3-030-41395-8_3

could have been discussed with no more than fleeting reference to the part played by governments. Today, few indeed are the sectors of such co-operation that do not require some sort of governmental action." The emergence of "science policy" in the national and international policy debates had been accompanied by a proliferation of intergovernmental organizations promoting or conducting scientific research. This was partly due to the change in the scope and nature of scientific activity, involving "the growing complexity and increasing specialization of the sciences, the accelerating rate at which discoveries are perfected and exploited, and the mounting cost of the equipment essential to research and of the research process itself."[2] The arrival of "big science" conducted in teams and requiring highly specialized expertise and know-how while swallowing increasing sums of money created a new rationale for sharing resources in an international framework. "Big science" projects were mostly also very visible, which guaranteed their attraction in the eyes of political decision-makers and the public.[3] But in Europe the interest in cooperation also was a legacy of the wartime collaborative projects between the allies, which had demonstrated the advantages that states could derive from joint efforts in research and development. Therefore, in contrast to the years before World War II, when international scientific cooperation had merely been concerned with irregular exchanges between individual scientists, cooperation now became a far more important and organized affair,[4] mostly involving large-scale projects, stimulated and financed by national governments. For the EC, entering this spawning field signified both challenge and opportunity.

During 1964, the European Commission, which until 1967 consisted of three separate executives: The Euratom Commission, the EEC Commission, and the High Authority of the European Coal and Steel Community (ECSC), played an active part in the various intergovernmental meetings on research policy, convened by the OECD. Moreover, the Brussels institution went along with the emerging concept of science as an essential vehicle for economic growth. This is clear, for instance, in the statement of the Euratom's representative Emmanuel Sassen in the OECD ministerial conference of October 1963: "[T]he European Economic Community, as well as EURATOM, have a keen interest in science policy as an element of economic growth. This interest is based in particular on the commitment of Member States under the Treaty of Rome to co-ordinate their general economic policies."[5] Sassen also told the participants of the meeting about the recent recommendation of the EEC Commission to the Council concerning medium-term economic policy. That document underlined the need to increase productivity, both due to demographic reasons and because of the imperatives of international industrial competition, which was increasingly reliant on the technical quality of fabrications. In addition, he thought that since public powers in all countries to an increasing extent intervened to stimulate scientific research and technical progress, it would be useful if these efforts were coordinated at the Community level.[6]

The OECD provided inspiration to consider broader policy competences for the EC in the field of research, but the organization was also a useful context for discussing, promoting, and testing the issue. From the first meeting of the OECD Interim Committee, which was set up as a result of the 1963 ministerial meeting, and where also the Commission was invited to participate,[7] Euratom's representative Gérard Louis de Milly reported: "It seems to me…necessary to draw the Commission's attention to the fact that during private conversations, several delegates of the Six [EC member states] in this meeting appeared very interested in an activity in this domain in the context of the Six."[8] In his view, this could offer the Commission an opportunity to present a concrete initiative on science policy.[9]

The idea of broadening the Community competences in the field of research was not entirely new. Some individual Commission officials had actually considered the issue as early as in 1958. An anonymous note of the Euratom Commission of July 1958 called for an overall approach to research policy:

> The treaties establishing the two Communities, the ECSC and Euratom, contain special provisions for scientific and technical research, but the problem of research as a whole has not yet been the subject of any disposition within the Community of the six countries. … There is, obviously, a gap that would be most useful to fill, and which should be the subject of a thorough review. … Indeed, scientific and technical research is one of the most powerful factors contributing to maintaining and raising the standard of living.[10]

Interestingly, the essential theme of the 1960s debate was already there: the Community needed a general research policy because of the impact research was perceived to have on the standard of living. No immediate action, however, followed from this note that seemed to have been forgotten among the piles of papers of the Brussels administration.

In 1960, the three Executives organized a conference "Technical Progress and Common Market." Even though the participants—professors, scientists, government experts, and "persons who regularly deal with these problems and/or bear particular responsibility in these fields, especially employers organisations and trade unions"[11]—recognized the interdependence between the Common market and technical progress, few still drew the direct line between science and growth or spoke of "science policy" or "research policy" as a necessity in international economic competition. The main concern of that meeting was clearly the negative social consequences of technical change.[12] But the Vice-President of the Commission of the EEC, Robert Marjolin, who also was a professor of economy and the first secretary-general of OECD, elaborated somewhat different ideas. "The first condition of the Common Market is rapid economic progress and the first condition for rapid economic progress is rapid technical progress," Marjolin stated in his presentation. "But above all, the Common Market must enable the economies of our six countries to benefit fully from the technical progress and to exploit all possibilities. It is because the national markets are too small, and because the development of national conjunctures is too uncertain, that the Common Market has been realized to allow the optimal use of all the possibilities to increase production."[13] Marjolin, however, was not very explicit in what fully benefiting from technological progress and exploiting all possibilities within the Common Market would eventually entail. Should the Community be given new competences? If so, what could these competences be?

Comparing the Commission's considerations soon after 1963 and the discussion in the 1960 Conference, it is notable how much the intellectual framework had changed. The impact of the OECD debate is evident: now it was the concern of growth, not the negative effects of technology that dominated the discussion, while "research policy" had emerged as a central concept and an explicit political objective. This change also attracted the attention of the Euratom Commission's President Pierre Chatenet, who in November 1963 wrote a letter to Dino Del Bo, the president of the High Authority of the ECSC. Referring to the OECD ministerial conference of the previous month, Chatenet noted that in Europe the problem of interdependence between scientific development and economic growth not only occupied an increasingly important place in public debates but also was devoted more attention in political milieus, national governments, and international organizations. Even though the Treaties of Rome and Paris had not given the European Communities direct competence to promote research in general, they included relative competences of development and cooperation of research in specific sectors as well as competences in the economic and social fields. Therefore, the work started in the Conference "Technical Process and Common Market" of

1960 should be continued. "I have the impression," Chatenet wrote, "that the time has come to take up this problem, especially as the European Parliament has repeatedly insisted on the need for scientific and technical coordination at a European level and on the development of a genuine science policy of the Communities."[14] Consequently, Chatenet proposed the creation of a study group bringing together the three Executives to examine the interdependence of the development of science and economic growth on the one hand and the action of the Communities and research coordination on the other.[15]

Chatenet's proposal of an Inter-Executive Working Group coincided with the efforts of the EEC Commission to set up a Medium-Term Economic Policy Committee. A brainchild of Robert Marjolin, the Committee, composed of representatives both of the Commission and the national governments, was to establish five-year programs exposing the main lines of the economic policies that the Community institutions and the member states would follow.[16] More generally, the Committee intended to study questions related to scientific research.[17] These plans effectively diminished the EEC Commission's interest in the creation of a new separate forum to discuss research policy as suggested by Chatenet, which again had the unintended and paradoxical effect that the debate on research policy was blocked altogether. The Commissions of Euratom and the High Authority of the ECSC feared that they would be excluded from the works the EEC was about to start and entered a bitter disagreement with the EEC Commission.[18] Locked into their internal acrimonies, the three Executives were unable to advance the topic that all of them considered important. The initiative on research policy, which just within a couple of years had gained a unique momentum across Western Europe, was about to be stalled within the Brussels bureaucracy.

It is interesting that the essential push for EC research policy did not come from Brussels but from a member state, where the issue had been devoted particular political attention.[19] Two months after the start of the OECD Interim Committee's work, the French Research Minister Gaston Palewski wrote a letter to French Foreign Minister Maurice Couve de Murville, which contained an interesting proposal. Palewski pointed out that according to the Rome Treaty, the member states of the European Community should soon be in the position to arrive at a common economic and commercial policy. Underlining the importance of scientific research to economic growth, Palewski remarked that the Community did not yet have proper policy competence in this field, and argued that this situation should be changed: "It is clear today that the Six cannot achieve their goals without a common research policy."[20] This is why the French government and its partners should ask the Commission and the Council to define the first elements of a common policy of scientific research and technology and to determine how the Treaties could be modified to permit new activities.[21] Couve de Murville was responsive to Palewski's suggestion. However, he thought that since the Community institutions were already planning to study the possibility of launching common activities in the field, it would be more appropriate to follow the progress of these works and to concentrate on defining the French position toward them.[22]

Yet, Paris did not want to wait too long. In January 1965 France assumed the presidency of the EEC, and its work program for the period entailed a study on the elaboration of a common policy in scientific research and technology. On March 4 the French government submitted a memorandum to the secretary-general of the Council to be transmitted to the member states and the Commissions of Euratom and the EEC. The document, repeating that the objectives of the Rome Treaty contained harmonious development of economic activities and continuing and balanced expansion, contended that "these notions imply the necessity to make joint efforts in the field of scientific and technical research."[23] The EC thus needed a research policy in order to obtain its major political goal, which was economic growth.

Besides the rhetoric of growth and competitiveness, the OECD idea of a gap was firmly in place in France's proposal. Since the beginning of the decade, the document argued, Europe had no longer been catching up on its technological lag in relation to the United States. Therefore the risk of seeing the technology gap to widen between the two continents had become "very real."[24] For closing the gap, the French proposed two series of studies to the Community: the first would have the objective of comparing national research programs in the civil sector, while the second would concentrate on charting those industrial sectors in which efforts in applied research were clearly insufficient compared to other countries, and sectors whose dynamism conditioned the general development of scientific and technological research. The results of these studies could serve as a basis for the definition of a "real common research policy."[25]

At a first sight, the fact that it was France making this initiative might seem odd. Since 1963, European questions had been less of a priority for President Charles de Gaulle, who was striving for independence in a wider international arena. The French interest in the European project had also been fading as a result of the cooling of Franco-German relations following the departure of Chancellor Konrad Adenauer. De Gaulle disliked the Atlantic orientation of Adenauer's successor's Ludwig Erhard that did not quite fit his grand design for Europe. The French leader aimed at a gradual reduction of the American power on the continent and the creation of a "European Europe," a freestanding continental Western European bloc.[26] Furthermore, the Gaullist government appeared less and less tolerant of the Commission's growing self-confidence as well as to some of its manifest intentions to increase the political power of the Brussels institutions.[27] However, there are multiple good reasons for the French administration to suggest a Community-level research policy. First of all, in 1965, of the six EC members, France was clearly the one with the most developed and centralized structures of a research policy geared toward national economic objectives. In France, the first national institutions aimed at supporting research were set up already in the interwar period,[28] although the essential shift came only with the government of Pierre Mendès France (1954–1955) and the explicit interest of Mendès France to include research in national decision-making. Mendès France, contrary to most of his contemporaries, conceived of research policy as a political domain in its own right and wanted to make science into a political objective, a field of action and political reasoning.[29] His short time in office did not allow Mendès France to fully implement his plans, but after leaving government, he continued to promote the idea of a greater government role in science. For instance, Mendès France was closely involved in the organization of the Conference of Caen in 1956, which has commonly been perceived as the crucial event setting the premises for France's postwar research policy.[30]

The Fifth Republic and President de Gaulle continued the postwar push for research policy. The re-establishment of the Ministry of Science as well as the creation of new powerful institutions such as the Comité interministériel de la recherche scientifique et technique (CIRST), the Comité consultatif de la recherche scientifique et technique (CCRST), and the Délégation générale à la recherche scientifique et technique (DGRST),[31] all demonstrated decisive attempts by the government to develop centralized structures for a national science policy. In addition to the Sputnik Shock of 1957 and the increasing international tensions of the Cold War, the political shift was encouraged by a generational change: the turn of the decade saw the arrival to power of a new generation of scientists with fresh visions of the role of science in society.[32]

There were fundamental differences between the thinking of Pierre Mendès France and Charles de Gaulle: while the former saw research primarily as an element of his program to modernize the French economy and society, the latter, although equally engaged in the mission of modernization, regarded research more as a means to restore and secure French national independence and grandeur.[33] To a large extent, the "grand research policy" of de Gaulle, however, was based on and inspired by the ideas developed in the circles of Mendès France and the discussions in the Conference of Caen.[34] These inserted science into the postwar French project of modernization and growth that essentially would be realized through national planning. Scientific research was clearly associated with growth in the Third Plan (1957–1961),[35] while the Fourth Plan (1962–1965) explicitly placed research "on the top of the hierarchy of the factors determining the development of the nation"[36] and considered it as a "means of economic development."[37]

Indeed, when Paris launched its proposal for an EC research policy, the government was strongly steering the national research policy toward the practical goals of industry and economy. In 1965, the OECD noted that France was "giving very deliberate emphasis to technical research to economic ends, and particularly to the strengthening of specifically industrial research."[38] The French proposal of March 1965 was only a part of a broader political agenda aimed at launching a European industrial policy. Laurent Warlouzet has shown how, after 1964, some powerful figures in the French administration tried to reorient France's European policy from the strong focus on agriculture toward initiatives designed to bolster European competitiveness in the increasingly open world markets. According to Warlouzet, these individuals belonged to a new generation of political elite more oriented toward business, increasingly critical of France's traditional policy line in Brussels and sensitive to the imperatives of international competition.[39]

From the perspective of Paris, it was the EEC framework that would be best suited to the purpose of strengthening the European hand in research. The reasons for this preference were listed in the March 1965 proposal: the economic solidarity of its members, the specificity of its objectives, and the traditional vocation of Europe to provide the world with reputable scientists and scholars made the Community the most appropriate context for "vast and systematic" action.[40] But there was also another incentive for France to emphasize the Community context: in the EC the French were vital players who had the will and the means to steer Community activity toward primarily French interests. The OECD was vast, and it involved members who were too powerful to be controlled by France. After the 1961 reform, it was no longer a purely European enterprise but with the US membership now included a stronger transatlantic dimension.

The Gaullist opposition to supranationalism in Brussels was also never absolute. It is well known that in some areas where yielding power to supranational institutions was deemed necessary for promoting French interests, especially in fields such as common agricultural policy, Paris had little hesitation in allying with the Commission in order to secure its own agenda.[41] Moreover, in the 1960s, France was, as N. Piers Ludlow aptly has noted, doing extremely well from its European involvement; the French economy had benefited from the EEC liberation regime, and the French exports to the other Community member states were growing faster than any other of its partners, except Italy.[42] So, while critical about federalist traits of integration in general, the French must have seen that the Community was offering genuine possibilities that could work for their benefit.

The proposal of March 1965 was, indeed, not even the first attempt made by the French government to initiate research cooperation within the Six. Research had been included already in General de Gaulle's blueprints of political cooperation in 1960 and 1961. The idea was first presented at the summit of the Heads of State and Foreign Ministers in Paris in February 1961,[43] and as a consequence a Study Committee was set up including also a subcommittee (the "Pescatore Committee"), with the mission of examining the questions of culture and research.[44] Even though the Six welcomed the subcommittee's work,[45] there was no follow-up on this matter.[46] Cooperation in the field of science and culture was still included in the draft treaty that the French presented in October 1961 (Fouchet Plan I),[47] but the Pescatore Committee's work was not continued.[48]

There are, nonetheless, two important remarks to be made here. First, in the Fouchet Plan as well as in the works of the Pescatore Committee, research was envisaged as a part of strictly intergovernmental cooperation. No plans were made to grant the Community new policy competence in this field. On the contrary, with his 1961 proposal of political union, de Gaulle wished to absorb the Communities and thereby eradicate the inconvenient federal tendencies of European integration.[49] Although the memorandum of March 1965 cannot be perceived as a major concession to a supranational Europe on the part of the French, one can still note that the proposal was tightly situated within the Community framework. In 1968, Robert Gilpin argued that in terms of integration, here Paris was ready to go quite far: "What the French have in mind in calling for a European science policy is not merely cooperation in science and technology but eventually a common policy toward American economic policies and – especially – investments."[50]

Second, in the initiatives of 1960 and 1961, research was perceived as auxiliary to cultural and educational policy, a framing entirely different from the memorandum of March 1965.[51] The emphasis on growth and competitiveness was not present in these considerations that exclusively focused on fundamental research, whose objectives were rather cultural than economic. In 1965, economy was at the center of the French overture. This shows how much the ideational context for research policy had changed in Europe and France just within a few years.

The French proposal gave important support to the Commission's endeavor to extend the Community's activities in the field of research. After being discussed in the Committee of Permanent Representatives (COREPER, a Council body of national delegates) in mid-March 1965, the initiative was passed to the Commission of the EEC with a mandate to study it and report to the Council by the following October.[52] Meanwhile, the research policy debate had resumed in Brussels. The Medium-Term Economic Policy Committee had discussed research for the first time about a month before the COREPER meeting (with several delegations mentioning the close relationship between science and economy).[53] In April, there was an agreement to establish a special working group "to study the problems of developing a coordinated or common policy for scientific and technical research, and to propose measures for initiating such a policy, taking into account possibilities of co-operation with other countries."[54] The mandate of the group Politique de la recherche scientifique et technique (PREST) also mentioned the close link that these works would have with more general economic objectives.[55] PREST was more or less what the French had wanted with their memorandum. In its meeting on June 14, 1965, the group launched a number of studies on the conceptions and policies of the EEC countries in the domain of research and technology as well as on problems that research policy could be expected to resolve in the following years.[56] The choice of a Frenchman, André Maréchal, the head of the DGRST, to chair PREST only underlined the French connection.

The PREST members were high government officials who were responsible for science policy and represented their own countries. The German members of PREST, for instance, came either from the finance ministry or the research ministry. France, on the other hand, was represented by officials from the DGRST, the CCRST, or the Commissariat général du Plan, while most Italians in PREST held positions in the national research council, Consiglio Nazionale delle Richerche. There were also two Commission members in the working group: an official from the DG Economic Affairs and a Special Councilor.[57] Similar to the Medium-Term Economic Policy Committee, PREST was thus a strictly intergovernmental forum with almost no direct representation from the science community.

PREST was not the only institutional innovation of 1965 that demonstrated the Community's new interest in research. The Inter-Executive Working Group that Pierre Chatenet had proposed in 1963 was finally set up in October 1965.[58] That group soon launched studies on the evolution and coordination of research activities in the Community.[59] Yet, all these works proceeded slowly. By the following June, PREST had a record of three meetings only: in addition to its opening meeting, the working group had gathered twice to produce the interim report. The number of the meetings of the Inter-Executive Working Group was not much higher: by October 1966 there had only been three gatherings.[60] PREST published a first version of its final report in July 1967.[61] Mostly, the delay was due to the drastically deteriorated political climate among the six Community member states. With their "empty chair" policy,[62] the French had abandoned their seats in Brussels, a move that practically paralyzed the Community for half a year. On the other hand, the Germans were, still in 1967, unprepared to take a clear stand in the issue. Wolfram Langer, the president of the Medium-Term Economic Policy Committee, received explicit advice from his government to try to limit the discussion on the new forms and areas of European cooperation in research as long as the German interministerial working group, set up in the autumn 1966 on the initiative of research minister Gerhard Stoltenberg, had not completed its work. Important tensions also prevailed between the German finance ministry and the research ministry.[63]

Yet the PREST report, as the first official intergovernmental Community document systematically exploring the opportunities for a common research policy, offered a kind of milestone in these early debates. Moreover, it presented an ideational framework where science policy was rather straightforwardly subjected to the objective of economic growth:

> In order to ensure the continued growth of the economy and to control the effects of scientific and technological progress, research efforts in all Member States must be increased. Economic and social progress in the coming decades depends mainly on progress in research and development.[64]

Moreover, although PREST openly recognized the difficulty in determining, in quantitative terms, the relationship between science and growth, this difficulty was only thought to stem from the novelty of the domain and the consequent lack of information.[65] There was also a clear tendency in the Medium-Term Economic Policy Committee to orient PREST even more toward economic thinking.[66]

Overall, the working group followed the line of discussions taking place in the OECD. Besides the stress on the importance of science for economic expansion, the concern about the technology gap was strongly present in the documents that the working group delivered. The interim report contained careful calculations of research spending in the EC countries and comparisons of the figures between the EC, the United States, and Great Britain—simultaneously revealing striking divergences, not only between the Community and the two non-member countries but also within the Six. According to the figures in the report, in 1962–1963, the Americans spent 3.1 percent of their GNP on research and the British 2.2 percent, the equivalent French amount being 1.6 percent and the German 13 per cent. The numbers of the other EC member states were smaller still. The report also stressed the utility of the international comparisons that for the time being constituted the only criteria for evaluation and orientation of government efforts in research. Furthermore, references to other countries had an additional advantage in the way they "make it possible to take a step towards taking into account scientific research as an increasingly important factor in international competition."[67] PREST clearly embraced research as an economic asset increasingly necessary for success in the international markets.

The remarkable similarity of the concepts and ideas of the PREST reports to those in the documents originating from the OECD and the French circles, has an easy explanation: many of the people involved in these different institutions were the same. André Maréchal, the president of PREST, represented France in the OECD meetings on science. Pierre Cognard, the head of Service du Plan and the French person who allegedly coined the concept of the "gap" was not only closely involved in the OECD but also chaired meetings of the PREST deputy members.[68]

Maréchal and Cognard were also not the only individuals spanning boundaries between the French administration, the OECD, and the EC, and thereby facilitating the flux of political ideas. There were many others, such as Pierre Piganiol, the head of the DGRST, an influential author of OECD reports, and a strong supporter of a Community research policy. In an article, published in *Minerva* in 1968, he underlined the need to consider scientific organization at the EC level, and concluded that "[t]he research efforts of the Six should be directed towards narrowing the 'gap' which exists between the Community and the United States." Piganiol also proposed the creation of a European Science Foundation, an idea that after various discussions was finally realized in 1974.[69] Louis Villecourt, advisor to Piganiol at the DGRST, who went to work at the OECD Directorate for Scientific Affairs while continuing as a part-time adviser to Piganiol's successor André Marèchal, used very similar language. Villecourt ended up in the EEC, which the British observers thought was a result of his "personal frustration with the OECD, when his ideas on studies of 'the technology gap' found disfavour with the Americans."[70]

The PREST report made a strong case for a common EC research policy, arguing that the Community should be considered as a "privileged platform" and the problems of cooperation arising from cooperation between the EC countries and third countries should be studied in a way that would not hurt the Community interest.[71] It also included more specific proposals, such as the creation of a European system of collection and diffusion of scientific information, the harmonization of fiscal regimes that also should be made favorable to investments supporting innovation, an agreement on a European patent and European enterprise, and the elaboration of a coherent doctrine of political intervention both at the member state level and at the Community level, the regular comparison of national plans, programs, and budgets in order to arrive at *concertation* of national policies, and finally, the establishment of common programs. In addition to the grand domains of activity, namely, nuclear, space, and aviation, the report named six priority sectors of eventual cooperation: information technology and telecommunications, transport, oceanography, metallurgy, nuisances, and meteorology[72]—all fields that in the mid-1960s promised rapid leaps of technical progress and a positive impact on industry.

By 1967, the discussion on Community research policy had thus been transformed from abstract debates among OECD influencers and some Commission officials into concrete proposals with clearly defined purpose:

boosting European economic performance against the American rivals. It was time to bring the issue to the member governments that ultimately would make the decision of any new Community activity. This happened in the very first EC Council meeting devoted to research on October 31, 1967. On this occasion, the governments of the six member states gave the EC the green light to develop new activities in this domain. In the final communiqué of the meeting, the ministers expressed "Their willingness to implement, in connection with the medium-term economic development program of the Community and taking into account new developments in the field of research, a vigorous action of recovery and promotion of scientific and technical research and of industrial innovation."[73]

Overall, the meeting was perceived as a success. German diplomatic reports spoke of "true willingness" to reach a positive outcome,[74] and even hailed the event as a "beginning of a new phase of European cooperation."[75] Indeed, the outcome took many contemporary observers by surprise. For instance, the British foreign ministry that closely followed the events in Brussels, predicted the meeting would be "rather a damp squib," not least because "[t]he countries of the Six are obviously either bereft of policy and ideas of science and technology, or completely unable to agree on the ideas put up to them (notably by the French)." For the British, aspiring for EC membership, the prospects of the Council looked so dim that they even considered making profit from it. A note from the Foreign Office speculated five days prior to the meeting: "If our own ideas had been any better developed, this could have been an ideal time to come forward with them."[76]

There was fortuitous mix of converging interests, strategic alignment, and shared ideational ground behind the outcome of the Luxembourg Council meeting. Perhaps the most decisive contribution to the Council preparations came on the part of the Commission. The Commission, although at times divided, had been an early and rather systematic supporter of the idea of granting the Community new activities in research. Sure enough, the fusion of the three Executives in 1967 complicated matters: during that year, the Commission was unable to work to its full capacity.[77] On the other hand, the establishment of the General Directorate of general research (DG Research) in July 1967 created a venue where most of the Commission's so far scattered efforts in this domain could be concentrated. This not only improved the efficiency of the Brussels institution but also formed an institutional framework that later guaranteed continuity in the Community's activity in the field. A seemingly and deliberately minor institutional reform that was basically unobserved at the time, it came to have important long-term consequences for European politics.

Crucial too is that the new Commission could rely on the work of its predecessors. An important legacy of the High Authority and the old Commissions of the EEC and Euratom was the memorandum "on the problems of the scientific and technological progress in the European Community," submitted to the Council in March 1967. That document, where the three Executives presented their reflections on the desirable orientations of the Community's activities in research, formed an important basis for the discussions in Luxembourg.[78] Moreover, after the humiliation of the Empty Chair Crisis in 1965, there was a burning need for a positive overture: the crisis had generated an overall sense that rapid evolution of the EEC had become impossible.[79] Visible development in a new policy field appeared to the Commission as an extraordinary chance to enhance its profile and show that against all odds, progress could be made. Besides, research policy might also have been part of the Commission's strategy to meet the challenge of the Gaullist France by binding the country more tightly to the EC. With this logic, enmeshing France in an increasingly complex set of economic and political ties would make it difficult for de Gaulle to strike out at the Community. In the best case, research could have worked like agriculture, and be a field of entente and mutual interest between the Commission and Paris, guaranteeing that the French would not carry out their threatened attack against the common institutions.[80]

A second important factor behind the accord in the Luxembourg Council was the activity of the Italian and Belgian governments. In fact, it was the on the initiative of the Belgian Foreign Minister Pierre Harmel of December 1966, that the special Council meeting was organized.[81] The relative weakness of Belgian research effort made it reasonable to look for stronger partners able to share their resources. After Luxemburg, Belgium was the second smallest member state in the Community, and it only had very limited possibilities to conduct large-scale research on its own. Moreover, the strong concentration of Belgian industrial potential in traditional sectors such as coal mining had retarded the development in newer, more research-intensive industries. According to the OECD country report on the organization of scientific research of 1963, Belgium had "few very large firms in the most recently developed branches of industry, which relies primarily on research." This was in addition to the nearly complete absence of the defense sector and the fact that Belgian industry was seldom asked to undertake research on government account.[82]

The Italians also had pressing concerns about the current status of European—and their domestic—science and technology. In the summer 1966, the Italian government expressed in the NATO Council its anxieties concerning the growing technological differences within the Alliance and proposed closer cooperation in research. The two Italian memoranda that circulated among the NATO partners envisaged the establishment of a structure or vehicle comparable to the Marshall Plan, with the objective of eliminating the technology gap within a period of ten years. Under this design, the United States would provide the technical knowledge that Europe was lacking, while the European governments would finance joint development projects in specific sectors (such as computers, aviation, space research, satellites, and atomic energy).[83] Important here is that the proposals, further elaborated upon during the following autumn, not only contained the idea of fortifying the Atlantic partnership but, by referring to the earlier French initiative, also suggested an increase in European collaboration. Moreover, the European Community, owing to its perceived role as the core of European unification, was presented by the Italians as a favorable basis for joint continental effort.[84]

Antonio Varsori has argued that the reasons for the proposal were first and foremost political. In 1966, anti-American sentiments were strengthening in France. This occupied the Italian government, keen to secure the continued presence of the United States in Europe and worried about France's destabilizing acts which threatened the political development within the Community of the Six. Suggesting cooperation with the United States in a field where the French had shown strong interest, the Italians hoped to improve the mutual relations of these two countries. Furthermore, Italy's quest for a "European relaunch" matches the general tendency of the increasing activism of Italy in European affairs since the early 1960s, prompted by not only the change of government but also the tightening ties between France and Germany. The rapprochement of the two big continental countries provoked fears of a powerful Franco-German bloc within the EC where Italy would find itself marginalized.[85] It is also likely that Rome's overture was motivated by the wish to secure a bigger share from the Community's budget, heavily dominated by sectors of the common agricultural policy that brought little benefit to Italy.[86]

But behind the Italian initiative, there was also a genuine concern about Italian research. In the 1960s there were only a few sectors, like nuclear physics, where world-class research was conducted in Italian institutions, while Italy's figures of state-funded R&D remained clearly behind its European partners.[87] Despite the initiatives launched by the Christian Democrat and Socialist Party government in the early 1960s to encourage Italian scientific effort, the tendency among the country's ruling class had been to treat research as an issue of secondary concern.[88] By the mid of the decade, however, there was an increasing awareness of the relative backwardness of Italian science and technology,[89] and the Italian government started to show interest in international cooperation with technologically more powerful countries.[90] This went hand in hand with the strong association of the European Community with *miracolo economico*, Italy's impressive economic development after the war.[91]

The Italian proposal sparked off a vivid discussion about the ideal context for joint European research. Although the Italians emphasized the merits of the EC, the initiative was nevertheless presented in a NATO meeting and addressed to all NATO countries. This was no accident: especially after 1957 there had been considerable efforts to use the organization to strengthen Western European science. The creation of NATO's Science Committee "to speak authoritatively on science policy" and to guarantee "the full development of our science and technology," which was seen as "essential to the culture, to the economic and to the political and military strength of the Atlantic community,"[92] was not just a panicked reaction to the Sputnik shock. The nuclear stalemate between the superpowers and the following observation that an essential part of the Cold War conflict had shifted onto non-military terrain, constituted an indispensable background to the establishment of the Committee.[93] Consequently, the organization's research activity focused less on military research and more on fostering scientific achievement more generally. This was done mainly through two educational programs: a fellowship program that was aimed primarily at the international exchange of postgraduate and post-doctoral students, and a program devoted to the organization of summer schools and the support of advanced study institutes. According to John Krige, besides the emphasis on basic, non-military research, the appointment of eminent scientists to the Science Committee guaranteed NATO high legitimacy in the eyes of the Western scientific community. For the very same reason, international cooperation within the NATO framework was politically feasible: far from direct application to military objectives, it did not threaten national defense interests.[94] These experiences, together with the strong American presence in the organization, were good reasons for Fanfani to select the Alliance to the political context in which to present his proposal.

But whatever NATO's role and reputation in the field of research, the reference to the Atlantic framework in the Italian proposals was widely seen as problematic. During the discussion in the Euratom Council of December 6, 1966, the President of the EEC Commission, Walter Hallstein, while greeting the Italian declaration as an important contribution to the ongoing work on the definition of a common policy of technology and science, stressed the need for the Community to define its own policy before seeking solutions in a wider context.[95] The option of using NATO as an institutional basis for developing European scientific and technological cooperation was later also rejected by the Germans who deemed the organization, due to the military character of its objectives, as inappropriate for that purpose. Since the creation of a totally new organization was also undesirable, cooperation in research and technology had to be primarily realized within the Six. Unsurprisingly, the French, traditionally critical of organizations with strong American presence, adopted a similar stance.[96] But even the British, who were elaborating their own plans for European technological cooperation, pressed Fanfani on the need to look away from the transatlantic context.[97]

This clear lack of enthusiasm among its partners about NATO-based cooperation led Rome to shift its focus and support the proposal that the Belgian government had made to discuss the issue within the EC Council. In the spring of 1967, they joined the Belgians in stressing the need to organize a special Council meeting as soon as possible.[98] These two governments also submitted the Council Secretariat notes containing guidelines for the elaboration of a common policy in research.[99] Moreover, Belgium and Italy had an influence on the outcome of the Council through their efforts to temper the staunch resistance of the Dutch government, which mainly derived from the issue of the British membership.[100] Winning the Dutch consent was crucial for the agreement in Luxembourg.

The renewed British interest in EC membership since the spring of 1966, and especially the official membership application of May 1967, had made the Community enlargement a topical question again.[101] Since the Netherlands was a strong supporter of the British accession to the EC,[102] the Hague did not want to engage with developing new Community activities in which Britain was not offered the possibility to participate from the outset. Consequently, the Dutch, using a variety of arguments, attempted to slow down the planning on the meeting of the Community's research ministers.[103] In Luxembourg, they also presented a draft resolution that envisaged postponing the discussion until late 1968 at least. In the face of the pressure of Italian and Belgian governments, the Netherlands finally agreed to compromise.[104]

But more than threatening to delay decisions on EC research policy, the question of enlargement sparked discussion on the scope and reach of the Community's future activities. The way in which the British government combined technology and its objective of the entry to the Community, gave an important boost to the plans to increase joint effort in research. Most manifestly this thrust came with Prime Minister Harold Wilson's suggestion of a "European Technological Community" (ETC) in late-1966. The overture was directly related to the revision of the British European policy and the increasing interest of the Labour government in EC membership. Undoubtedly, the move was inspired by the advances made by Foreign Minister Fanfani. Foreign Office records reveal how London wanted to make "the best possible use of European anxieties about technological backwardness, and of the Italian initiative, as a lever of our approach to Europe."[105]

Wilson had good reasons to believe that mentioning the possibility of technological cooperation could attract the EC countries, especially France, to perceive UK accession to the Community in more favorable terms. After World War II, Britain had been the only European country able to resume scientific research almost without a break. It was also in Britain where in the immediate postwar years the government paid most attention to the matter.[106] With its R&D spending significantly higher than in any capitalist country other than the United States, Britain was clearly the scientific and technological leader of Western Europe. Moreover, since the early 1960s, research gained even more political visibility. "The extraordinary wave of political excitement about science and technological revolution" had been a crucial factor sweeping the Labour Party into power. Between 1959 and 1964, science was highly politicized and placed at the forefront of the 1964 election campaign.[107] A concrete manifestation of this trend was the establishment of the Ministry of Technology (Mintech), described by its head, Anthony Wedgewood Benn, as "the first techno-economic ministry in the world."[108] The high political profile of science, together with the image of Britain as a European leader in the field, transformed science and technology into a precious political trump card which the Labour government was keen to use.

While the Wilson administration approach to the Fanfani initiatives was highly strategic,[109] European cooperation was also perceived in London as a true chance to compete with the United States in advanced technologies.[110] In London, a clear difference was made between the "propagandist presentation of the 'technological community' theme" and the "actual formulation of proposals to be put to the Six."[111] Fundamentally, as for the Italians, as also for the British, the Community offered the primary platform for activity. Only the rationale for this emphasis was different. From the British point of view, the weak competitiveness of Europe's science-based industries compared to those of the United States resulted less from the lack of European technological know-how than from the environment in which European firms were operating. In the first place, the strength of the US industry derived from the ability of basing its operations in a very large home market, and therefore the best way to proceed in Europe would be enlarging the European Community and creating the necessary marketplace in Europe.[112]

Ultimately, Wilson's strategy to use science and technology to enter the European Community failed. The main target of Wilson's strategy, the French President de Gaulle, remained unmoved.[113] Predictably, de Gaulle questioned the sincerity of the British overture and suspected that in the end London would still prefer cooperation with the United States. The fate of the Labour government's tactics was sealed by the second French veto of the British membership in December 1967. In that context there was little help with the purposeful ambiguity that the Wilson administration preserved around the initiative to protect it from French attacks.[114] The proposal's vagueness, together with the simultaneous signs of British disengagement from some major existing projects of European technological cooperation, provoked skepticism and confusion on the continent, even among those supporting UK membership.[115]

For the Six, Wilson's overture, however, constituted an important stimulus to pursue the Community's work on research. The British proposals, after all, explicitly placed the discussion on European scientific and technological cooperation in the EC framework. A quote from the speaking notes on technological collaboration which Wilson was using during his visits to the European capitals in 1967 shows this approach rather clearly: "I conceive the European technological community as an integral part of the enlarged European Economic Community, with the United Kingdom as a member. There is and will continue to be technological interchange between Britain and the other European countries, but this cannot achieve its full effectiveness and cannot become a positive element in our industrial policies unless we abolish the tariff and other barriers to trade which presently exist between us."[116] Cooperation in technology and enlarged liberalized markets, thus, went hand in hand. It is also interesting that the prospect of establishing a new separate ETC, or fostering integration in a context other than the Community, was rejected outright by most EC member governments. In December 1966 Sir Solly Zuckerman, the scientific adviser to the British government, noted after having visited the Netherlands and France: "In both places I was asked whether the Prime Minister's reference to a 'Technological Community' meant that we were going to propose another international organisation – the implication being that any such proposal might be regarded as 'empty' by the people with whom we would wish to cooperate."[117]

The complex political context of the Luxembourg EC Council discussions also included the evolution of the German position on the proposals laid out by France, the Commission, the UK, and the others. Germany had been slow to formulate its official position on research policy, and consequently, until 1967 there were no major initiatives coming from the federal government. Also, the tone of Bonn's documents for the research Council meeting remained cautious.[118] As the fall progressed, however, there was a progressive change in Germany's approach, which largely was stimulated by the responsibilities of the Council Presidency.[119] One can also perceive a broader interest in international research cooperation in the federal government. Indeed, the new Foreign Minister Willy Brandt envisaged science and technology as an important element of foreign policy, especially for a country like Western Germany that for political and historical reasons had to rely on other means of influence than military capacity. Furthermore, Brandt thought that in addition to preserving peace, guaranteeing the East-West détente and diminishing the differences between industrialized and developing countries, scientific and technological cooperation could be used to reinforce the process of European integration.[120] These thoughts were clearly expressed in Brandt's radio speech of August 1967,[121] and they were echoed in a statement of Research Minister Stoltenberg, who a week after the Luxembourg Council meeting said that the EC should be made the fulcrum of European research policy.[122]

This change reflected a broader adjustment in German de-centralized research policy. The unexpected recession of 1966/1967 had triggered the federal government to significantly increase its political and financial engagement with research. Now the promotion of research and technological innovation was openly recognized as an instrument of economic policy that would help draw the country out of recession, while the concerns of the technology gap were given more political attention. The second *Bundesbericht* on research of 1967 suggested a drastic rise of government support for research up to 3 percent of the GDP.[123]

Consequently, Germany made a decisive contribution to the preparations for the Luxembourg meeting. Fritz Hellwig, the German Commissioner responsible for research, recalled that cooperation between the Commission and the German Ministry of Science was crucial for bringing the PREST report to the Council on time.[124] Having a German as the head of the newly established DG Research was crucial. Hellwig was able to create contacts with the administration of his own country and keep the federal government informed of the Commission's work.[125] Germany also openly supported both the Italian[126] and British initiatives on research.[127]

The legacy of France in guaranteeing the success of the Luxembourg Council meeting and thus pushing research policy into the Community's agenda is far more mixed. During the charged months of the empty chair crisis, the French government lost its immediate interest in the plans of a common research policy and European scientific cooperation more generally. A concrete manifestation of this was the absence of the French in the PREST meetings for almost one year. But this was also the spirit in statements of the French representatives in the COREPER meetings during early 1967,[128] and in President Pompidou's declaration at the Franco-German summit of January 1967. According to Pompidou, the organization of European research cooperation in the framework of the Six was difficult. This was not only because of the importance of the participation of Britain in these activities (France had vetoed the UK entry only a month before) but also due to the fact that none of the EC member states apart from France and Germany was making particularly strong efforts in the field.[129] He concluded bluntly that "[i]f we want to do something other than just to express nice words, we all know that it will essentially be about German and French technology."[130] Indeed, France was now driving fast on the bilateral line: in 1966, Chancellor Erhard and French Prime Minister Pompidou had agreed to organize, within the framework of the 1963 Elysée Treaty on Franco-German cooperation, periodical consultations between the research ministers of both countries.[131] There were also more concrete initiatives, such as the creation of the Institut

Laue-Langevin on fundamental research on neutron, the first major joint research project between France and Germany, in January 1967,[132] which was soon followed by the signature of an intergovernmental convention on the development of a common telecommunications satellite, another exclusively Franco-German undertaking in the field of science and technology.[133] In Luxembourg, France adopted a conciliatory approach, but it stemmed from strategic calculations: Paris wanted to show to Britain how easily the cooperation just within the Six could be organized even in the fields that were not covered by the Treaties, and that the UK did not need to join the EC to benefit from this work.[134] Two days after the Council, Minister Schumann's cabinet director told a member of the British Foreign Office that the French were very pleased with the Luxembourg decisions and their potential political use. By welcoming non-member countries to participate in future discussions on EC research policy, France had brilliantly foiled the British attempt to play the technology card.[135]

The activity of the European Commission and the Italian and Belgian governments, together with the encouraging signs coming from London, and the constructive approach of Germany—and finally even France—constituted the particular political context that facilitated the outcome of the Luxembourg Council in October 1967. But one more element should still be added to this complex picture: the agreement in Luxembourg was essentially eased by the evolution of political discourse and the consolidation of a specific policy-framing for research, stressing growth and competitiveness. By 1967, the OECD discussion on science policy had already born fruit. The second OECD ministerial meeting in 1966 demonstrated how the issues deliberated in the same forum only three years earlier had now become common knowledge across Western Europe. Research was commonly perceived as a necessary part of government policy and an essential tool to economic growth. In the 1966 ministerial conference, the chairman of the OECD interim Committee on science policy noted: "This interdependence [between science and economic growth] was now generally conceded, even though its detailed mechanisms could not be clearly discerned. Nearly all countries now had institutions, differing in their organisational forms, tendencies and terms of reference, which were shaping or carrying out science policies at the national level."[136] No longer was science policy a foreign concept to the Western European governments. It had become an obligation and an important political goal.

The year 1967 was also the time when the debate of technology gap reached its zenith. The 40,000 copies of Jean-Jacques Servan-Schreiber's book *Le défi Américain* sold within a few months after its publication, illustrating the responsiveness of the European public to the idea of American scientific and technological challenge. Against this backdrop it is not very surprising that the gap was also at the heart of the discussions in the Luxembourg Council and found its way to the final communiqué of the meeting, stating that:

> developments made in recent years by European countries in science, technology and industrial applications have been slower than those observed outside Europe, mainly in the United States, in a number of industries essential for the development of modern industrial economies; and that Europe's delay in this area creates a serious risk for its economic and social development in the medium and long term.[137]

By 1967, the technology gap, thus, had developed into an officially shared idea of a risk factor threatening European economic development.

Especially sensitive to the dangers of the gap seemed to be the Belgians. The note that the Belgian government submitted to the Council on October 13, 1967, affirmed that due to the difficult position of the European industry vis-à-vis the United States, a common activity in the Community appeared "indispensable and urgent."[138] Furthermore, the Belgian Prime Minister van den Boeynants declared in the Council of October 31 that in case of failure to "closely unite" its forces "in time," Europe would fulfill the vision of decline.[139] Revealing is also how the Belgians expressed their uneasiness about the way the British used the expression of technology gap. "[I]t would be very helpful to the Belgians," noted a cable from the UK Embassy in Brussels, "if we could refrain from prefixing the phrase technological gap with the epithet 'so-called'. It might well be that in a number of fields there was no real gap between the British and United States technology, but this was not the case in Belgium, which was in many respects lagging very badly."[140]

The concerns of the gap were also strongly embraced by the Italians. A document the Italian government circulated to its EC partners in late 1966 started with the affirmation that:

> the technological problems of the West and the gap between certain countries, especially the United States, are a matter of serious concern and also require, a political initiative that will promote continuous and balanced progress of the countries belonging to the Atlantic zone.[141]

Engaging in common action was even more important because the time was perceived to be running out. The language of urgency was amply employed throughout the paper, speaking of the "augmentation of retard," and a difference in production capacity that was "more and more pronounced." Moreover, besides the economic consequences of the gap, the Italians highlighted the political and psychological impact European delay could have in the Cold War competition. A serious economic weakening in the Atlantic zone, they thought, would diminish the attraction of the Western European camp vis-à-vis the opposing bloc.[142] This resort to Cold War rhetoric, otherwise rare in the gap debate, was not exceptional in Italian declarations. At the Luxembourg meeting the Italian Research Minister Leopoldo Rubinacci not only stressed the necessity and urgency to define a "European technological program" but also repeated that the objective of the Soviet Union was to accelerate the application in traditional civilian economic activities of technology and know-how emanating from military and space research.[143]

Neither within the Commission nor in the member governments of the European Community was the gap undisputed. In the Community debates the gap regularly emerged as a complex phenomenon that the actors were trying to understand and define. One important disagreement concerned the actual location and the extent of the gap. In which sectors could Europe be said to be in retard? Were some countries in a worse position than the others? In the Luxembourg Council meeting, the Belgian Prime Minister, Paul van den Boeynants, identified US advance with regard to three issues: the development of technology in the most advanced domains, the volume of public investments supporting research and industrial innovation, and the development of management methods in both public and private sectors. In his view, the American progress in these fields had been clearly more significant in the last ten years, which had placed Europe in a position of relative inferiority. The economic consequences of this situation were already being felt. The balance of licenses and patents was negative for Europe, as was also the case with trade, with the Europeans exporting the less profitable products of traditional industrial sectors. American investments—especially in high technology—in the European markets were also rapidly increasing. Grégoire, the Luxembourg foreign minister, on the other hand, rather than stressing the American might, underlined the factors that he thought accounted to European weakness: the insufficient investment in research, the lack of dynamism in industrial application, the emigration of researchers, and finally, the incapacity of the small European countries.[144] The diagnoses and their degree of alarmism varied from panic-infused statements of an overall dispersion and weakness of European research and innovation to less emotional analyses (mostly by the British), reminding that while a transatlantic difference in productivity and standard of living existed, "[i]n the last 15 years, economic growth has been faster in Western Europe than in U.S.A.; the difference in standard of living has narrowed; and Western Europe[an] exports have increased much faster than American."[145]

Despite the heterogeneity of statements and beliefs, the gap played a vital role in the Luxembourg Council meeting. The Foreign Office even attributed the success in Luxembourg to the technology gap, stating that "the feeling of a need to make a common effort to wipe out the American technological lead carried the day over individual considerations. Would it have been the same had there been less consciousness of the risks implied by the overbearing American superiority or had it been possible to believe more firmly in the benefits of close cooperation with the British?"[146] By 1967, the gap had become so powerful factor in the debate on EC research policy that it shaped the decisions of European ministers.

Essentially, the idea of a gap was based on a specific perception of science and its role in industrialized societies: European technological weakness signified economic weakness. Technology, produced through scientific research, was necessary for national and European growth and performance in the international markets. It is, nonetheless, interesting that the connection between scientific research and economic growth was not taken for granted in all interventions during the 1967 debate. For instance, the memorandum of the three Executives of March 1967 highlighted that none of the numerous studies devoted to the examination of this linkage "does not allow to establish a precise relationship between the two phenomena. The existence of this relationship may even be questioned in view of certain empirical observations."[147] The experience showed that there were countries which, despite their remarkable success in research and technology, had only mediocre growth rates, whereas some countries that had been investing less in research were growing faster. "It is because the link between research and growth is indirect," the document concluded, "that it cannot be perceived statistically."[148]

Furthermore, growth was not always identified as the only objective of research policy. The same memorandum noted that "the goal of growth is far from being the only one, not even often the main foundation for the increased efforts of all advanced nations in research. Through growth and alongside it, research must also serve social purposes whose importance is fundamental in the framework of the Community as well as for the ends of the cultural and political order."[149] Nevertheless, the economic aspect remained paramount. The draft resolution, attached to the memorandum and for this part accepted by the Council of October 31, 1967, without modifications, included clear language: "Progress in science and technology is a fundamental factor in the economic growth and the general development of the Member States of the Communities, and in particular in their competitive capacity."[150] The narrow interpretation of research policy, emphasizing competitiveness and growth, won out in Luxembourg.

This political framing is apparent in the interventions of the national delegations at the meeting. German Research Minister Gerhard Stoltenberg said in his opening speech that since the EC certainly constituted one of the closest forms of existing international economic cooperation, it *naturally* also had to deal with the problems of research policy. Therefore, increasing importance should be given to the intimate connection between research, industrial development, and economic expansion. The other German delegate, Hans von Heppe, though adding that the significance of research was not limited to its economic relevance,[151] pointed out that since research and development conditioned the economic expansion and the competitiveness of enterprises, it was necessary for the Community to widen its existing cooperation into new domains. By the same token, the Belgian Prime Minister Paul van den Boeynants announced that in his government's view, cooperation, which would be limited to research and development without more concrete industrial and commercial objectives, constituted a trap for Europe. He was followed by the Dutch Economic Minister Leo de Block who supposed that scientific research and its technological applications had fundamental importance for economic development. It is also illustrative that the opinion of Luxembourg's Foreign Minister Pierre Grégoire that common activities in research should not be restricted to the technological domain and his warnings that material preoccupations could lead to an increasing "dehumanisation" of the European civilization almost seemed out of place among the many declarations stressing the importance of research for economic growth and competitiveness.[152]

In order to understand this discursive landscape dominated by worries of losing out in international economic rivalries, one has to gauge the changes in the broader international context. The late 1960s was marked by the calming of international military tensions. The diminution of the prospect of a direct armed confrontation between the United States and the Soviet Union gave more room to economic rivalries that were not necessarily fought along the traditional Cold War lines. Moreover, the rapprochement of the two superpowers aroused worries among the Europeans of being marginalized in international affairs, while the increased European sense of self-esteem following the years of economic boom altered the bonds with the United States. All these developments had a specific impact on the process of European integration.[153]

For European discussions on research, in particular, the change in the transatlantic relations was important. In the course of the 1960s, the United States appeared as a less and less reliable ally. This was clearly manifested in the space sector where Americans were assuming an increasingly unilateral course. In September 1965, President Johnson approved a restrictive policy concerning US assistance in the development of the communication capacity of foreign satellites. Although in the following July that approach was revised, the policy remained restrictive.[154] Similar trends were visible also in the nuclear sector. When the Community sought in March 1967 to double the quota for plutonium imports from America, the United States replied by making the agreement conditional on a tighter inspection system, a further exchange of information, and a much higher price.[155] With a growing distance from the Americans, and the international economic and technological competition sharpening, Europeans seemed to have no other choice than to combine their forces.

But there was also another important international factor that had an impact on the discussions on European research policy: the changing global division of labor. In the Luxembourg Council meeting, this point was raised by Commissioner Fritz Hellwig, who emphasized the consequences that the changing production structure in the developing countries would have for Europe. The improving capacity of newly industrialized regions to compete in the sectors traditionally dominated by European enterprises threatened Europe's position in world markets and forced them to make structural reforms that would essentially involve higher investment in research and technology. Hellwig also remarked that these reforms were possible only in an expanding economy, and that, therefore, ensuring continuous growth—which again had its source in research-intensive industries—had to remain a compelling political objective of the Community.[156]

These changes went hand in hand in the first signs of a new industrial revolution, the arrival of which Jean-Jacques Salomon observed in 1967:

"The whole question is whether all the investments made in research and development do not lead to a new industrial revolution of which the United States would be both a source and a model."[157] This entailed a redefinition of the industrial societies that were now becoming "scientific societies," or "knowledge societies" not only applying science to the organization of industrial production but also organizing themselves with regard to scientific and technological production. In the competition between the "scientific societies," propelled by rapid US-led trade liberalization, the scope and rapidity of innovation were increasingly crucial for success. This is something many observers thought the individual European countries could not realize alone.

The Council meeting of October 31, 1967, was the culmination of over a year's debate. Besides the general approval to the idea of developing new activities within the Community, the meeting also had some very concrete results: now there was an agreement on the content and a time-table of the Community's future work in the field of research. The working group PREST was charged to examine the possibilities of cooperation in the sectors identified in its report (information technology and telecommunications, transport, oceanography, metallurgy, pollution, and meteorology) as well as to chart the possibilities of collaboration in other sectors. Furthermore, the Council gave the working group the task of comparing national research activities and studying the possibilities to create a common system of diffusion of technical information. In the future, PREST would examine means to advance coordination in education and to increase international exchange of scientists. A reference in the communiqué to the possibility of third countries to participate in the activities opened prospects for cooperation beyond the Community framework. The Council asked PREST to submit a new report before March 1968. Likewise, the COREPER, assisted by a group of high officials and in close association to the Commission, was instructed to present its conclusions by the following June.[158]

While certainly marking a milestone, the significance of the very first EC Research Council meeting should not be exaggerated. A striking feature of the Council minutes and surely also a further factor behind the agreement was the abstract level of the discussion. The suggestions of the different national delegations concerning future activity remained very vague and differed widely. The only common denominator seemed to be the creation of an economic climate favorable for research and innovation that would include measures such as the introduction of a European patent and European commercial enterprise. Otherwise the proposals varied. While the Italians wanted to improve the education of scientific personnel and the diffusion of knowledge, and envisaged the creation of common research centers, the Luxembourg and German delegations advocated comparative studies of national activities, which would enable the definition of priority actions and the optimal institutional framework for eventual cooperation. The Luxembourg delegation also stressed that the coordination of activities should be stretched to other international organizations, such as the OECD where they hoped the Six could unite their so far different positions. Most ambitious proposals came from the Belgian delegation that spoke of "integrated cooperation" at least in the "key sectors" of advanced technology. But even the Belgians remained vague on that point. According to Van den Boeynants, that cooperation would only be aimed at organizing public actions in the member states and their industrial potentialities in a way that enabled concrete realizations responding to common needs. Moreover, these prospects clearly did not please the French Minister Schuman who said that France did not want the Community to have supranational means of economic intervention. The Dutch delegation also warned of the isolation of the Community if the cooperation could not be realized in a framework larger than the Six.[159]

Finally, the countries did not commit themselves to significant action; in the final analysis, the communiqué's mention of an "energetic action of redressing and promoting scientific research and industrial innovation" remained a statement imprecise enough to satisfy all. Not even the concept of "science policy" was included in the final resolution. The British, in particular, were quite blunt on this. A Foreign Office cable from Brussels to London stated:

> It does not seem that this first council meeting to discuss technology has achieved a great deal. The resolution adopted and the programme contained in it, deal more with the questions of procedure than those of substance. The first subjects chosen for intensive study seem a surprising mixture of the important and the relatively peripheral. Most of the really difficult decisions, which will determine whether a community policy on technology comes to life, remain to be taken.

At the same time, this was already more than they expected: "in view of the political background against which the meeting took place, it is perhaps surprising that this was decided."[160]

Clearly, the outcome was not enough for the Commission, which hoped that the EC would create a research policy with far-reaching cooperation and coordination in a number of new sectors. Hans Michaelis, the head of the cabinet of Fritz Hellwig, described the resolution as "meagre" and "too limited," and therefore as an insufficient basis for a common technology policy.[161] Some Commission officials were also not very optimistic about the follow-up. In November 1967, Michel Albert, the French director in the Commission responsible for the medium-term economic policy and an active advocate of integration,[162] told the British that he was extremely gloomy about the mandate approved by the Council. Even if at the official level some proposals certainly were worked out and even saw the light of the day, they most probably would be scotched by the political rivalries between the member states. If not, the Dutch blocked them in case no progress was being made on the British application for membership, and then they would be dumped by the French who perhaps regarded them as too ambitious and supranational.[163]

But still, the Luxembourg meeting had an indisputable symbolic significance. Already the fact that the Community's ministers responsible for research had come to the meeting and seriously discussed the topic together, served as a clear demonstration that two and a half years after the submission of the French proposal on a common research policy, the Community was recognized worthy of consideration as the framework for action. Even the skeptical British paid attention to the symbolic meaning of the Luxembourg outcome and the prospects it eventually opened: "It is regarded in Community circles as a first step, albeit very small one in a field which may come to absorb an increasing amount of the Community's attention and effort in the coming years and which is seen by many as of great importance for the eventual success of the Community."[164] Luxembourg was an important opening that prepared the ground for the events that followed. It also showed how just within a few years, the debate on science policy as an engine for economic growth had been absorbed by the European Community and turned into concrete political proposals. Powered by the fears of a technology gap and the unsatisfiable quest for greater economic expansion, the idea of an EC research policy started to take shape and influence political agendas.

This is remarkable bearing in mind the novelty of not only research policy but scientific cooperation in Europe more generally. Although the postwar years witnessed an unprecedented number of international large-scale government-run projects in the field of science and technology, in many ways, this was a virgin field. Pierre Piganiol, the author of the legendary 1963 OECD science policy report, wrote in 1968 that "[i]n some countries, they [research institutions] have not yet defined their tasks, lack administrators capable of understanding, forming and directing them and have not yet discovered ways of collaborating with other research institutions."[165] His observation is supported by the findings that Albert H. Teich made while interviewing and surveying scientists in joint European research laboratories in the mid-1960s. Out of almost 300 respondents, only about 9 percent reported having studied at a higher educational institution outside of their own country, and for almost half of those with that experience, the host country had been the United States. "Here, among perhaps the most internationally-oriented group of scientists in Europe, one finds strong evidence of a lack of scientific interchange among European universities," Teich concluded.[166] It took long to change this. Still in 1976, Pierre Aigrain, the French Délégué général à la recherche scientifique (1968–1973) and State Secretary responsible for research (1978–1981), was able to proclaim: "The French have a higher chance to encounter a German scientist in an American laboratory than a German laboratory, and that is a shame."[167] The idea of a common research policy between the six European Community countries was set to break down these barriers and accelerate integration in a field with newly found political significance. Despite the initial excitement, it turned out to be a bumpy ride.

NOTES

1. The Treaty of Rome, 25 March 1957. http://ec.europa.eu/economy_finance/emu_history/documents/treaties/rometreaty2.pdf (accessed January 26, 2019).
2. Jean-Jacques Salomon. "International Scientific Organizations," in *Ministers Talk About Science*, 57, 64–65.
3. Helmut Trischler and Hans Weinberger, "Engineering Europe: Big Technologies and Military Systems in the Making of 20th Century Europe," *History and Technology*, 21 (2005): 64.
4. Salomon "International Scientific Organizations," 57, 64–65; Corine Defrance and Anne Kwaschik, "Sciences, internationalisation et guerre froide. Éléments d'introduction." In *La guerre froide et l'internationalisation des sciences. Acteurs, réseaux et institutions*, ed. Corine Defrance and Anne Kwaschik (Paris: CNRS Editions 2016):

11–28; Naomi Oreskes and John Krige Krige, eds. *Science and Technology in the Global Cold War* (Cambridge, M.A.: The MIT Press, 2014).

5. The author's translation from French.

6. OCDE: Conférence Ministérielle sur la Science. Procès-verbal de la Conférence réunie au Château de la Muette, à Paris le jeudi 3 octobre et le vendredi 4 octobre 1963. BAC 3/1978-786, Historical Archives of the European Union (hereafter HAEU).

7. Euratom (Commission, Direction Générale des Relations Extérieures): Compte rendu de la 1ère réunion du Comité Intérimaire créé par la Conférence Ministérielle sur la Science, 3–4 octobre 1963, Paris, Château de la Muette, les 13–14 janvier 1964. BAC 3/1978-786, HAEU.

8. The author's translation from French.

9. Euratom, Relations extérieures: Note à l'attention de la Commission. Conférence Ministérielle sur la Science Paris, le 3–4 octobre 1963. Possibilités d'action pour Euratom, Bruxelles, le 21 janvier 1964. BAC 3/1978-786, HAEU.

10. Quoted in Arthe van Laer, "Vers une politique de recherche commune. Du silence du Traité CEE au titre de l'Acte Unique," in *Les trajectoires de l'innovation technologique et la construction européenne = Trends in technological innovation and the European construction*, eds. Cristophe Bouneau, David Burigana, and Antonio Varsori (Brussels and New York: P.I.E. Peter Lang, 2010): 23. The author's translation from French.

11. European Conference "Technical Progress and the Common Market." BAC 21/1966-140, HAEU.

12. Exposé de synthèse effectué par Monsieur Paul Finet, Membre de la Haute Autorité à la Conférence Européenne Progrès Technique et Marché Commun, Bruxelles, le 10 décembre 1960. BAC 001/1971-2, HAEU.

13. Exposé général de M. Robert Marjolin, Vice-Président de la Commission de la CEE, Bruxelles, le 6 décembre 1960. BAC 001/1971-2, HAEU. The author's translation from French.

14. The author's translation from French.

15. Pierre Chatenet to Dino Del Bo, November 13, 1963. BAC 118/1986-1399, HAEU.

16. Laurent Warlouzet, *Le choix de la CEE par la France. L'Europe économique en débat de Mendès France à de Gaulle (1955–1969)* (Paris: Comité pour l'histoire économique et financière de la France, 2011), 349–350, 354–355.

17. Van Laer, "Vers une politique de recherche commune," 34; Ausschuß für mittelfristige Wirtschaftspolitik. Probleme der Forschungspolitik (Sekretariatsdokument). BAC 27/1985-6, HAEU.

18. Groupe Interexecutif Recherche, U. Zito to G. Cangellario, October 26, 1966. BAC 118/1986-1399, HAEU.

19. The Commission had already noted that in the first OECD Interim Committee meeting, of all Community's member states, the most enthusiastic about the idea of increasing the EC's activities in research was clearly France. Euratom, Relations extérieures. Note à l'attention de la Commission. Conférence Ministérielle sur la Science Paris, le 3–4 octobre 1963. Possibilités d'action pour Euratom, Bruxelles, le 21 janvier 1964. BAC 3/1978-786, HAEU.

20. The author's translation from French.

21. Le Ministre d'État chargé de la Recherche Scientifique et des Questions Atomiques et Spatiales à Monsieur le Ministre des Affaires étrangères, Paris, le 3 mars 1964. Europe: Coopération économique 1961–1966, 1487, Archives du ministère des Affaires étrangères, Paris Courneuve (hereafter AMAE).

22. Le Ministre des Affaires étrangères à Monsieur le Ministre d'État chargé de la Recherche Scientifique et des Questions Atomiques et Spatiales, le 28 mars 1964. Europe: Coopération économique 1961–1966, 1487, AMAE.

23. Note du gouvernement français sur l'élaboration d'une politique commune de la recherche scientifique et technique, le 4 mars 1965, Bruxelles, le 9 mars 1965. CM2 1967-1029, HAEU. The author's translation from French.

24. Ibid.

25. Ibid.

26. Mark Trachtenberg has questioned the consistency of de Gaulle's design and suggested that in reality the president's rhetoric different from his actual policy. Mark Trachtenberg, "The de Gaulle Problem." *Journal of Cold War Studies*, 14 (2012): 81–92.

27. Jean-Marie Palayret, "De Gaulle Challenges the Community. France, the Empty Chair Crisis and the Luxemburg Compromise." In *Visions, Votes, and Vetoes: the Empty Chair Crisis and the Luxembourg Compromise Forty Years On*, ed. Jean-Marie Palayret, Helen Wallace, and Pascaline Winand (Brussels and New York: P.I.E. Peter Lang, 2006, 47–49); N. Piers Ludlow, *The European Community and the Crises of the 1960s: Negotiating the Gaullist Challenge* (London and New York: Routledge, 2006), 57.

28. Vincent Duclert, "Pierre Mendès France et la recherche scientifique. Le sens d'une action gouvernementale." In *Le gouvernement de la recherche. Histoire d'un engagement politique de Pierre Mendès France au général de Gaulle (1953–1969)*, eds. Alain Chatriot and Vincent Duclert (Paris: La Découverte, 2006), 23–24. According to Jean-Jacques Salomon, the interwar activity was partly inspired by the Soviet experience of linking science and politics. Jean-Jacques Salomon, "Science Policy Studies and the Development of Science Policy." In *Science, Technology and Society: A Cross-Disciplinary Perspective*, eds. Ina Spiegel-Rösing and Derek de Solla

Price (London: Sage, 1977): 47. This added to a long legacy of strong state intervention in scientific affairs stemming from the Napoleonic period. Peter J. Bowler, and Iwan Rhys Morus, *Making Modern Science. A Historical Survey* (Chicago and London: The University of Chicago Press, 2005), 330.

29. Duclert, "Pierre Mendès France," 45–46, 54–55.
30. Vincent Duclert, "Le colloque de Caen, second temps de l'engagement mendésiste," in *Le gouvernement de la recherche. Histoire d'un engagement politique de Pierre Mendès France au général de Gaulle (1953–1969)*, eds. Alain Chatriot and Vincent Duclert (Paris: La Découverte, 2006): 81–82.
31. OCDE: Conférence Ministérielle sur la Science 3 et 4 octobre 1963. Chapitre I de l'ordre du jour, politique scientifique nationale. Document particulier – France. Paris, le 3 octobre 1963. BAC 3/1978-786, HAEU.
32. Dominique Pestre and François Jacq, "Une recomposition de la recherche académique et industrielle en France dans l'après-guerre 1946–1970. Nouvelles pratiques formes d'organisation et conceptions politiques," *Sociologie du travail*, 38 (1996): 273–274.
33. Alain Chatriot and Vincent Duclert, "Une portée historique." In *Le gouvernement de la recherche. Histoire d'un engagement politique de Pierre Mendès France au général de Gaulle (1953–1969)*, eds. Alain Chatriot and Vincent Duclert (Paris: La Découverte, 2006): 184; Pierre Lelong, "Le général de Gaulle et la recherche en France," In *De Gaulle en son siècle. Actes des Journées internationales tenues à Unesco Paris, 19–24 novembre 1990.* Tome III *Moderniser la France* (Paris: Institut Charles de Gaulle, 1992): 649.
34. Jean-Louis Crémieux-Brilhac, Une politique pour la recherché, in *Le gouvernement de la recherche. Histoire d'un engagement politique de Pierre Mendès France au général de Gaulle (1953–1969)*, eds. Alain Chatriot and Vincent Duclert (Paris: La Découverte, 2006): 198.
35. Robert Frank, "The French Alternative: Economic Power through the Empire or through Europe?" In *Power in Europe II? Great Britain, France, Germany and Italy and the Origins of the EEC, 1952–1957*, ed. Ennio Di Nolfo (Berlin and New York: Walter de Gruyter, 1992): 162–163. See also: Jean-Louis Crémieux-Brilhac, "Pierre Mendès France, l'enseignement et la recherche." In *Pierre Mendès France et le mendéisme. L'expérience gouvernementale (1954–1955) et sa postérité*, eds. François Bédarida and Jean-Pierre Rioux (Paris: Fayard, 1985): 435–451.
36. Ghrislaine Bidault, Un paysage institutionnel transformé. Planification, régionalisation et organismes (1958–1985), in *Le gouvernement de la recherche. Histoire d'un engagement politique de Pierre Mendès France au général de Gaulle (1953–1969)*, eds. Alain Chatriot and Vincent Duclert (Paris: La Découverte, 2006): 150.

37. Le Ministre d'État Chargé de la Recherche Scientifique et des Questions Atomiques et Spatiales à Monsieur le Ministre des Affaires Etrangères, Paris, le 20 avril 1964. Europe: Coopération économique 1961–1966, 1487, AMAE.

38. OECD, *Ministers Talk About Science* (Paris: OECD, 1965): 39.

39. These included Jean Dromer, Francois-Xavier Ortoli, Jean Saint-Geours, Michael Albert and Alain Prate. Warlouzet, *Le choix de la CEE par la France*, 420, 428, 447.

40. Note du gouvernement français sur l'élaboration d'une politique commune de la recherche scientifique et technique, le 4 mars 1965. CM2 1967-1029, HAEU.

41. N. Piers Ludlow, "The Making of the CAP. Towards a Historical Analysis of the EU's First Major Policy," *Contemporary European History*, 14 (2005): 347–371.

42. Ludlow, *The European Community*, 53–55.

43. Pierre Gerbet, *La construction de l'Europe* (Paris: Imprimerie nationale, 1999), 239–240; Georges-Henri Soutou, "Le General de Gaulle et le plan Fouchet." In *De Gaulle en son siècle. Actes des Journées internationales tenues à Unesco Paris, 19–24 novembre 1990.* Tome V: *L'Europe* (Paris: La Documentation française – Plon 1992), 131–132.

44. Jean-Marie Palayret (with the assistance of Richard Schreurs), *A University for Europe, Prehistory of the European University Institute in Florence (1948–1976)* (Rome: Presidency of the Council of Ministers, Department of Information and Publishing, 1996): 111, 115–116.

45. Parlement Européen, commission de la recherche et de la culture. Projet d'avis de la commission de la recherche et de la culture à l'intention de la commission politique sur la proposition de résolution de M. Gaetano Martino, le 27 février 1967. CEAB2-2622, HAEU; Déclaration des chefs d'Etat ou de Gouvernement, Bonn, July 18, 1961. Direction Politique Europe, organisations internationales 1961, AMAE.

46. Parlement Européen, commission de la recherche et de la culture. Projet d'avis de la commission de la recherche et de la culture à l'intention de la commission politique sur la proposition de résolution de M. Gaetano Martino, le 27 février 27, 1967. CEAB2-2622, HAEU.

47. Gerbet, *La construction de l'Europe*, 241.

48. Anne Corbett, *Universities and the Europe of Knowledge. Ideas, Institutions and Policy Entrepreneurship in European Union Higher Education Policy 1955–2005* (Basingstoke and New York: Palgrave Macmillian, 2005), 50.

49. Soutou, "Le General de Gaulle et le plan Fouchet," 129, 137–138.

50. Robert Gilpin, *France in the Age of the Scientific State* (Princeton: Princeton University Press, 1968), 416.

51. Coopération en matière d'enseignement supérieur et de recherche. Rapport de la Commission des études aux chefs d'état ou de gouvernement, Paris, le 7 juillet 1961. Direction Politique Europe, organisations internationales 1961, AMAE.

52. It was also agreed that the work would be done in collaboration with the Euratom Commission, the High Authority of the ECSC as well as the EEC Medium-Term Economic Policy Committee. Comité des représentants permanents: Extrait du compte rendu sommaire de la réunion restreinte tenue à l'occasion de la 334ème réunion, Bruxelles, les 17, 18 et 19 mars 1965. CM2 1965-1029, HAEU.

53. Comité de politique économique à moyen terme. Réunion des Suppléants des 25.-26.2.1965. Résumé des conclusions. BAC 27/1985-3, HAEU; Projet de compte rendu succinct de la troisième réunion du Comité de Politique économique à moyen terme, le 5 mars 1965. BAC 27/1985-3, HAEU.

54. The author's translation from French.

55. Comité de politique économique à moyen terme. Mandat pour un groupe de travail "Politique de la Recherche Scientifique et Technique". CEAB2-3746, HAEU.

56. Comité de Politique Économique à Moyen Terme: Groupe de travail "Politique de la Recherche Scientifique et Technique," réunion du 14 Juin 1965, résumé des conclusions. CEAB-3748, HAEU.

57. Comité de politique économique à moyen terme: Projet de compte rendu succinct de la 4ème réunion du Comité de Politique Economique à Moyen Terme, le 9 avril 1965. BAC 118/1986-1395, HAEU; Francesco Bobba to Pierre Chatenet April 26, 1965. BAC 118/1986-1395, HAEU.

58. La Commission de la Communauté européenne de l'énergie atomique: Dixième rapport général sur l'activité de la Communauté (mars 1966–février 1967), 111. There was also a very informal subcommittee, chaired by the Secretary-General of Euratom Jules Guéron. Projet de compte rendu de la réunion du 14 octobre 1965 du Groupe Interexécutive de recherche scientifique et technique. CEAB2-3746/1, HAEU; Guazzugli-Marini to Wellenstein, March 15, 1967. BAC 118/1986-1399, HAEU.

59. Groupe de Travail Interexécutif, Recherche Scientifique et Technique. Projet de compte rendu de la réunion du Comite ad hoc du Groupe de travail interexécutif de la recherche du 23 novembre 1965. BAC 118/1986-1399, HAEU.

60. Groupe Interexecutif Recherche, U. Zito to G. Cangellario, October 26, 1966. BAC 118/1986-1399, HAEU.

61. Sommaire du rapport du groupe de travail "Politique de la recherche scientifique et technique." Pour une politique de la recherche et de l'innovation dans la Communauté. BAC 118/1986-1398, HAEU.

62. The immediate reason for the crisis was the failure to agree on the financial arrangement of the common agricultural policy as a result of which the French government withdrew its representatives from the Community institutions. Ultimately, however, the crisis stemmed from the clashing visions of the Community's future development and especially regarding the amount of political power that should be given to the Commission and the European Parliament. For a more in-depth analysis on these events, see for instance Palayret, Wallace, and Winand, *Visions, Votes, and Vetoes*.

63. Europäisches Zusammenarbeit auf dem Gebiet der Forschung und Entwicklung. Vermerk den Herrn AL I vom 7. November 1966, Bad Godesberg, den 21. November 1966. B 138/3944, Das Bundesarchive Koblenz (hereafter BAK); Peter Weingart and Niels C. Taubert. "Das Bundesministerium für Bildung und Forschung." In *Das Wissenschaftsministerium. Ein halbes Jahrhundert Forschungs- und Bildungspolitik in* Deutschland, eds. Peter Weingart and Niels C. Taubert (Weilerswist: Velbrück Wissenschaft, 2006), 14.

64. Comité de politique économique à moyen terme: Groupe de travail "Politique de la recherche scientifique et technique," Rapport intérimaire. BAC 118/1986-1396, HAEU. The author's translation from French.

65. Ibid.

66. Vermerk für Herrn von den Groeben, Brüssel, den 1. Juni 1966. BAC 062/1980-148, HAEU.

67. Comité de politique économique à moyen terme: Groupe de travail "Politique de la recherche scientifique et technique," Rapport intérimaire. BAC 118/1986-1396, HAEU. The author's translation from French.

68. Cognard first aired his worries about a "science gap" in an article published in *Progrès scientifique* in 1964. Julie Bouchard, *Commet le retard vient aux Français: analyse d'une rhétorique de la planification de la recherche 1940–1970* (Villeneuve d'Ascq: Presses de Universitaires du Septentrion, 2008): 96.

69. Pierre Piganiol, "Scientific Policy and the European Community," *Minerva*, 6 (1968), 359–361.

70. EEC Ministerial Meeting on Technology. Note of Alan Smith, October 25, 1967. PRO/FCO 55/59, The National Archives, London (hereafter TNA).

71. Sommaire du rapport du groupe de travail "Politique de la recherche scientifique et technique." Pour une politique de la recherche et de l'innovation dans la Communauté. BAC 118/1986-1398, HAEU.

72. Ibid.

73. Problèmes de la recherche scientifique et technique dans les Communautés. Annexe IV au doc. R/1893/67, CM2 1967-73, HAEU. The author's translation from French.

74. Der Bundesminister für Wissenschaftliche Forschung: Europäische Zusammenarbeit in Forschung und Entwicklung. Tagung des Ministerrats der Europäischen Gemeinschaften am 31. Oktober 1967, Bonn, den 5. Dezember 1967. B 136/5974, BAK.

75. Sachs aus Brüssel nach Bonn: 11. Ratstagung – to-pkt 3: Wissenschaftliche und technische Forschung in Europäischen Gemeinschaften, nr. 1977, den 2. November 1967. B 136/5974, BAK.

76. EEC Ministerial Meeting on Technology. Note of Alan Smith, October 25, 1967. PRO/FCO 55/59, TNA.

77. Katia Seidel, *The Process of Politics in Europe. The Rise of European Elites and Supranational Institutions* (London and New York: I.B Tauris Publishers, 2010), 176.

78. Mémorandum sur les problèmes que pose le progrès scientifique et technique dans la Communauté européenne. Communication de la Haute Autorité de la CECA et des Commissions de la CEE et de la CEEA aux Conseils, Bruxelles, le 20 mars 1967. BAC 118/1986-1391, HAEU.

79. Palayret, "De Gaulle Challenges the Community," 76.

80. Ludlow, *The European Community*, 34.

81. Procès-verbal de la réunion restreinte tenue à l'occasion de la 118ème session du Conseil de la Communauté Européenne de l'Energie Atomique, Bruxelles, les mardi 6 et mercredi 7 décembre 1966. CM2 1966-0081, HAEU. The idea of the Council meeting of ministers responsible for research was not Harmel's invention, however. Periodic meetings of ministers in whose competence research fell were envisaged already in the Bonn declaration of July 1961. Déclaration des chefs d'Etat ou de Gouvernement, Bonn, July 18, 1961. Direction Politique Europe, organisations internationales 1961 (38.1.1.), AMAE.

82. OECD: *Country Reports on the Organisation of Scientific Research, Belgium*. Paris 1963, 24.

83. NATO Ministerial Meeting, Paris 14–16 December, 1966, Brief No. 9. PRO/FCO 55/50, TNA; Le retard technologique de l'Europe et l'opportunité d'une collaboration internationale, Rome, le 6 septembre 1966. BAC 118/1986-1390, HAEU.

84. Procès-verbal de la réunion restreinte tenue à l'occasion de la 118ème session du Conseil de la Communauté Européenne de l'Energie Atomique, Bruxelles, les mardi 6 et mercredi 7 décembre 1966. CM2 1966-0081, HAEU; Le Retard technologique de l'Europe et l'opportunité d'une coopération internationale, Rome, le 6 septembre 1966. BAC 118/1986-1390, HAEU; Déclaration du Ministre italien des Affaires

étrangères M. Fanfani au Conseil de ministres de la C.E.E. des 6 et 7 décembre au sujet de l'initiative italienne en matière technologie. BAC 118/1986-1390, HAEU.

85. Antonio Varsori, *La Cenerentola d'Europa: l'Italia e l'integrazione europea dal 1946 ad oggi* (Soveria Mannelli, Catanzaro: Rubbettino, 2010): 182–183, 205–206.

86. Italy was a net looser in the CAP. Following the difficult negotiations between 1961 and 1964, a common market had been created for meat, milk, cereals, and rice—all products that Italy was actually importing. In this way, Italy, an important agricultural producer itself, ended up supporting the production of its wealthier partners. This bias obviously angered the political decision-makers in Rome and spurred them to look for other areas of integration from which Italy would draw greater benefits. Ibid., 189–192.

87. Mauro Capocci, "Politiche e istituzioni della scienza: dalla ricostruzione alla crisi," in Francesco Cassata and Claudio Pogliano, eds. *Storia d'Italia 26: Scienze e cultura dell'Italia unita* (Torino: Einaudi, 2011): 283–296.

88. Claudio Pogliano, "Images and Practices of Science in Post-War Italy," in *Science and Power: The Historical Foundations of Research Politics in Europe*, ed. Luca Guzzetti (Luxembourg: European Communities, 2000): 187, 191. See also: Claudio Pogliano, "Le culture scientifiche e technologiche," in *Storia dell'Italia repubblicana. La transformazione dell'Italia: sviluppo, e squilibri*. Vol. II/2 (Torino: Einaudi, 1995): 267–296.

89. The growing desperation and disquiet of some scientists and industrialists about Italy's situation was manifestly expressed in a meeting of the Federazione delle associazioni scientifiche e tecniche, in July 1967. This meeting marked an important occasion where the opinions, fears, and intentions of Italian big industry were fully revealed. The main worries of the participants included the increasingly negative balance of payments in technological products, the slow change of attitudes in industry, and the weak links between university and public research institutions, on the one hand, and private enterprises, on the other. This debate had a clear effect on Italian politicians: the following year saw the creation of a new fund (Imi, Istituto mobiliare italiano), designed to finance research in all fields except nuclear and aerospace. Pogliano, "Le culture scientifiche e technologiche," 598–599.

90. In November 1966, the Italian Ambassador in London sounded out the possibility of bilateral cooperation between the UK and Italy. Similar inquiries were also made in the American direction, which a year later resulted in a new agreement on bilateral cooperation. Varsori, *La Cenerentola d'Europa*, 207.

91. Ibid., 167, 203.
92. Quoted in Lorenza Sebesta, "Choosing its Own Way: European Cooperation in Space. Europe as a Third Way between Science's Universalism and US Hegemony?" *Journal of European Integration History*, 12 (2006): 37.
93. John Krige, "NATO and the Strengthening of Western Science in the Post-Sputnik Era." *Minerva*, (38) 2000: 106.
94. Ibid., 95–96, 106–107.
95. Procès-verbal de la réunion restreinte tenue à l'occasion de la 118ème session du Conseil de la Communauté Européenne de l'Energie Atomique, Bruxelles, les mardi 6 et mercredi 7 décembre 1966. CM2 1966-0081, HAEU.
96. Italienische Erklärung über technologischen Rückstand Europas, Brüssel, den 7. Dezember 1966. B136/5975, BAK; Comité des représentants permanents, projet de compte rendu sommaire de la 415ème réunion tenue à Bruxelles, les mercredi 22 et jeudi 23 février 1967. CM2 1967-1030, HAEU; Europäische Zusammenarbeit auf dem Gebiet vom Forschung und Entwicklung, Bad Godesberg, den 14. Februar 1967. B 138/3944, BAK.
97. John W. Young, "Technological Cooperation in Wilson's Strategy for EEC Treaty," in *Harold Wilson and European Integration: Britain's Second Application to Join the EEC*, ed. Oliver J. Daddow (London: Frank Cass, 2003): 102.
98. Aide-mémoire sur l'échange de vues intervenu le 9 juin 1967 dans le cadre du Comité des représentants permanents, Bruxelles, le 27 juin 1967. CM2 1967-1930, HAEU; Examen des résultats de la réunion du Comité des représentants permanents du 9 juin 1967. BAC 118/1986-1399, HAEU.
99. Préparation de la session du Conseil consacrée à la recherche scientifique et technique, Bruxelles, le 13 octobre 1967. CM2 1967-1032, HAEU; Préparation de la session du Conseil consacrée à la recherche scientifique et technique, Bruxelles, le 19 octobre 1967. CM 1967-1032, HAEU.
100. Der Bundesminister für Wissenschaftliche Forschung: Europäische Zusammenarbeit in Forschung und Entwicklung. Tagung des Ministerrats der Europäischen Gemeinschaften am 31. Oktober 1967, Bonn, den 5. Dezember 1967. B 136/5974, BAK.
101. Ludlow, *The European Community*, 126, 133–137.
102. There were various reasons for the Dutch support for British membership, such as the perceived importance of Britain as a counterweight to possible Franco-German attempts at dominating the Community and the fear of remaining "locked up" in a limited and protectionist continental block. Anjo Harryvan, *In Pursuit of Influence the Netherlands' European*

Policy during the Formative Years of the European Union, 1952–1973
(Brussels: P.I.E. Peter Lang, 2009), 174, 197.

103. Comité des représentants permanents, projet de compte rendu sommaire de la 415ᵉᵐᵉ réunion tenue à Bruxelles, les mercredi 22 et jeudi 23 février 1967. CM2 1967-1030, HAEU; Réunion du Comité des représentants permanents du 9 Juin 1967, Aide-mémoire June 9, 1967 From P. Duchateau to Guazzugli-Marini. CM2 1967-1030, HAEU; Aide-mémoire sur l'échange de vues intervenu le 9 juin 1967 dans le cadre du Comité des représentants permanents, Bruxelles, le 27 juillet 1967. CM2 1967-1930, HAEU; Aide-mémoire sur l'échange de vues intervenu le 23 juin 1967 dans le cadre du comité des représentants permanents, Bruxelles, le 30 juin 1967. CM2 1967-1030, HAEU; Sachs aus Brüssels nach Bonn: 442. Tagung Ständigen Vertreter 27. Oktober – to-Punkt 7: Vorbereitung der Ministerratstagung über wissenschaftliche und technische Forschung, den 28. Oktober 1967. B/138-3946, BAK.

104. Ministère des Affaires étrangères. Adresse à diplomatie Paris, Bruxelles, le 2 novembre 1967. DPE, Organisations internationales 2042, AMAE.

105. Young, "Technological Cooperation," 95–97, 103; NATO Ministerial Meeting, Paris December 14–16, 1966, Brief No. 9. PRO/FCO 55/50, TNA.

106. Pierre Auger, "Scientific Cooperation in Western Europe." *Minerva*, 1 (1962): 428.

107. Austen Albu, "Mr Harold Wilson's Scientific Revolution." *Minerva*, 8 (1970): 602.

108. David Edgerton, The 'White Heat' Revised: The British Government and Technology in the 1960s, in *Science and Power: The Historical Foundations of Research Politics in Europe*, ed. Luca Guzzetti (Luxembourg: The European Communities, 2000): 207–212, 221, 223, 236.

109. This becomes clear in the following Foreign Office briefing for the NATO ministerial meeting of December 1966: "The discussion which the Italians want to hold in NATO is…relevant to our main objective only to the extent that it enables us to whet European appetites for technological co-operation with us. Nothing can be agreed in NATO which will directly serve our purpose; though any engagements taken of the kind the Italians proposed could considerable weaken our hand." NATO Ministerial Meeting, Paris December 14–16, 1966, Brief No. 9. PRO/FCO 55/50, TNA.

110. Young "Technological Cooperation," 95–96, 110.

111. Note of T.W. Garvey, June 30, 1967. PRO/FCO 55/43, TNA.

112. NATO ministerial meeting, Paris 14–16 December 1966, Brief No. 9. PRO/FCO 55/50, TNA, Technological collaboration. T.W. Garvey to Mr. Wright, December 8, 1966. PRO/FCO 55/41, TNA.

113. Gérard Bossuat, "De Gaulle et la seconde candidature britannique aux Communautés européennes (1966–1969)." In *Crises and Compromises: The European Project 1963–1969*, ed. Wilfried Loth (Baden-Baden: Nomos, 2001), 518.

114. Young, "Technological Cooperation," 96, 99, 101, 105, 109, 110; G. Willan to Mr. Garvey, June 30, 1967. PRO/FCO 55/43, TNA.

115. Young, "Technological Cooperation," 96–99, 110–111.

116. P.Q. No. 1324. PRO/FCO 55/49, TNA; Visit of Italian Prime Minister to London, June 1967. European technological cooperation, Brief by the Foreign Office, June 30, 1967. PRO/FCO 55/43, TNA.

117. Solly Zuckerman to Richard Clarke, December 7, 1966. PRO/FCO 55/41, TNA.

118. Entwurf einer Entschließung des Rats der Europäischen Gemeinschaften den 11. Oktober 1967. CM2 1967, HAEU; Ergebnisniederschrift über die Ressortbesprechung vom 25. Oktober 1967 im BMwF betreffend Vorbereitung der Ministerratstagung am 31 Oktober 1967 über europäische Forschungspolitik. B 136/5975, BAK.

119. Research was also mentioned among the tasks of the German Council Presidency: Vertretung der Bundesrepublik Deutschland bei der Europäischen Wirtschaftsgemeinschaft und der Europäischen Atomgemeinschaft: Überschicht über die Aufgaben während der 4. deutschen Präsidentschaft vom 1.7.–31.12.1967, Brüssel, den 7. Juni 1967. B 20-200 (1.639), Das Politische Archive des Auswärtigen Amts (hereafter PA AA).

120. Ulrich Pfeil, "Une politique scientifique pour l'Europe? Recherche et technologie pendant les années Brandt." In *Willy Brandt et l'unité de l'Europe. De l'objectif de la paix aux solidarités* nécessaires, ed. Andreas Wilkens (Brussels: P.I.E. Peter Lang, 2011), 355–357.

121. Außenpolitische Nutzung des deutschen wissenschaftlich-technischen Potentials, Bonn, den 29. Januar 1968. B 138/3946, BAK.

122. Wissenschaftspolitik als Element der Außenpolitik. Vortrag vor der Deutschen Gesellschaft für Auswärtige Politik am 8. November 1967. I-626 (Stoltenberg) 133/3, Archiv für Christlich-Demokratische Politik, Sankt-Augustin (hereafter ACDP). Stoltenberg, a North German Protestant who had earned his doctoral degree in history at Kiel, indeed made an effort to appear as a strong minister in a relatively weak ministry, repeatedly at odds with the much stronger Ministry of Finance. The American journal *Science* even compared him to the powerful British Minister of Technology Antony Wedgewood Benn, because he had shown "the same sort talent for mastering his subject and flair for drama-tizing the importance of his ministry." John Walsh, "German Science Policy: Bund Shifts the Balance." *Science,* March 22, 1968.

123. Helmuth Trischler, "Die 'amerikanische Herausforderung' in den 'langen' siebziger Jahren: Konzeptionelle Überlegungen." In *Antworten auf die amerikanische Herausforderung. Forschung in der Bundesrepublik und der DDR in den 'langen' siebziger Jahren*, eds. Gerhard Ritter, Helmuth Trischler, and Margit Szöllösi-Janze (Frankfurt and New York: Campus, 1999), 12–13.

124. Fritz Hellwig, Interview by Wilfred Loth and Veronika Heyde, Bonn, June 3, 2004. European Commission Memories, European Oral History, HAEU.

125. Fritz Hellwig: Interview by Veera Mitzner, Bonn Bad-Godesberg, February 3, 2010.

126. Gespräch des Bundesministers Brandt mit den Italienischen Außenminister Fanfani in Rom, den. 5 Januar 1967. In *Akten zur Auswärtigen Politik der Bundesrepublik Deutschland* (hereafter AAPD) 1967, n°8 (München: R. Oldenburg, 1998); Gespräch des Bundesministers Brandt mit dem italienischen Außenminister Fanfani in Rom, den 5. Januar 1967, AAPD 1967, n°8.

127. Dem Herrn Bundeskanzler: Sieben-Punkte-Vorschlag Premierminister Wilsons für europäische technologische Zusammenarbeit, Bonn, den 22. Dezember. B 136/5974, BAK.

128. Comité des représentants permanents, projet de compte rendu sommaire de la 415ème réunion tenue à Bruxelles, les mercredi 22 et jeudi 23 février 1967. CM2 1967-1030, HAEU; Réunion du Comité des Représentants permanents du 9 Juin 1967. Aide-mémoire le 9 juin 1967 from P. Duchateau to Guazzugli-Marini. CM2 1967-1030, HAEU; Aide-mémoire sur l'échange de vues intervenu le 9 juin 1967 dans le cadre du Comité des représentants permanents, Bruxelles, le 27 juin 1967. CM2 1967-1930, HAEU.

129. Entretien de M. Pompidou avec le Chancelier Kiesinger, Paris, le 13 janvier 1967. MAEF-29, HAEU.

130. Entretiens franco-allemands. Entretien élargi du 13 janvier 1967, Paris. MAEF-29, HAEU. The author's translation from French.

131. An die Generaldirektion "Auswärtige Beziehungen" der Kommission der Europäischen Wirtschaftsgemeinschaft: Bilaterale deutsch-französische Regierungsgespräche – Ergebnisse des Besuches Bundeskanzler Erhard und den deutschen Ministern in Paris am 7./8. Februar, Brüssel, den 9. Februar 1966. I-659 von den Groeben 054/3, ACDP.

132. Dennis Guthleben, "L'Allemagne, fer de lance de la politique européenne du CNRS?" In *La construction d'un espace scientifique commun? La France, la RFA et l'Europe après le 'choc du Spoutnik,'* eds. Corine Defrance and Ulrich Pfeil. (Brussels: P.I.E. Peter Lang, 2012), 106.

133. Claude Carlier, Le programme franco-allemand des satellites de télécommunications 'Symphonie'. Réussite technologique, échec commercial et

consequences sur la politique spatiale européenne, in *La construction d'un espace scientifique commun? La France, la RFA et l'Europe après le 'choc du Spoutnik'*, eds. Corine Defrance and Ulrich Pfeil (Brussels: P.I.E. Peter Lang, 2012): 238.

134. Aufzeichnung: Tagung des Ministerrats der Europäischen Gemeinschaften am 31. Oktober 1967 über Zusammenarbeit auf dem Gebiet der wissenschaftlichen Forschung und technischen Entwicklung, Bonn, den 9. November 1967. B 136/5974, BAK.

135. EEC and technology: French reactions to Minister's meeting. Tel No. 1097 from Paris to Foreign Office, November 3, 1967. PRO/FCO 55/59, TNA.

136. Ministerial meeting on science. Draft Minutes of the 2nd meeting held at the Château de la Muette in Paris on Wednesday 12th July 1966. OECD-1385, HAEU.

137. Problèmes de la recherche scientifique et technique dans les Communautés, Annexe IV au doc. R/189/67. CM2 1967-73, HAEU. The author's translation from French.

138. Préparation de la session du Conseil consacrée à la recherche scientifique et technique, Bruxelles, le 13 octobre 1967. CM2 1967-1032, HAEU.

139. Procès-verbal de la réunion retraite tenue à l'occasion de la 11ème session du Conseil, Luxembourg, le mardi 31 octobre 1967. CM2 1967-73, HAEU.

140. Brussels to Foreign Office, January 5, 1967. PRO/FCO 55/41, TNA.

141. The author's translation from French.

142. Le retard technologique de l'Europe et l'opportunité d'une collaboration internationale, Rome, le 6 septembre 1966. BAC 118/1986-1390, HAEU.

143. Procès-verbal de la réunion retraité tenue à l'occasion de la 11ème session du Conseil, Luxembourg, le mardi 31 octobre 1967. CM2 1967-73, HAEU.

144. Ibid.

145. Technological collaboration with Europe, Ministry of Technology December 23, 1966. PRO/FCO 55/41, TNA.

146. EEC and technology: French reactions to Minister's meeting. Tel No. 1097 from Paris to Foreign Office, November 3, 1967. PRO/FCO 55/59, TNA.

147. The author's translation from French.

148. Mémorandum sur les problèmes que pose le progrès scientifique et technique dans la Communauté européenne. Communication de la Haute Autorité de la CECA et des Commission de la CEE et de la CEEA aux Conseils, Bruxelles, le 20 mars 1967. BAC 118/1986-1391, HAEU. The author's translation from French.

149. Ibid. The author's translation from French.

150. Problèmes de la recherche scientifique et technique dans les Communautés, Annexe IV au doc. R/189/67. CM2 1967-73, HAEU. The author's translation from French.

151. 11. Ratstagung: Wissenschaftliche und technische Forschung in Europäischen Gemeinschaften, Brüssel, den 2. November 1967 nr. 1977. B 136/5974, BAK.

152. Procès-verbal de la réunion restreinte tenue à l'occasion de la 11ème session du Conseil, Luxembourg, le mardi 31 octobre 1967. CM2 1967-73, HAEU.

153. Jean-Jacques Sorel (Jean-Jacques Salomon), "Le retard technologique de l'Europe," Esprit, 365 (1967): 759.

154. Carlier, "Le programme franco-allemand," 235.

155. Christopher Layton, European Advanced Technology. A Programme for Integration (London: George Allen & Unwin, 1969), 117.

156. Procès-verbal de la réunion restreinte tenue à l'occasion de la 11ème session du Conseil, Luxembourg, le mardi 31 octobre 1967. CM2 1967-73, HAEU.

157. Sorel (Salomon), "Le retard technologique de l'Europe," 775.

158. Problèmes de la recherche scientifique et technique dans les Communautés. Annexe IV au doc. R/1893/67, CM2 1967-73, HAEU.

159. Procès-verbal de la réunion restreinte tenue à l'occasion de la 11ème session du Conseil, Luxembourg, le mardi 31 octobre 1967. CM2 1967-73, HAEU.

160. Council of the European Communities: Technology. Telegram No. 338 from Brussels to Foreign Office, November 1, 1967. PRO/FCO 55/59, TNA.

161. Weiterbehandlung der Vorschläge der Gruppe Aigrain. Memorandum an Herrn Vizepräsident Dr. F. Hellwig von H. Michaelis, Brüssel, den 1. Oktober 1969. 386 Hellwig N/1359, BAK; Gipfelkonferenz. Memorandum von H. Michaelis an Herrn Vizepräsident Dr. F. Hellwig, Brüssel, den 30. Oktober 1969. K 169/ I-083 Hellwig, ACDP.

162. Warlouzet, Le choix de la CEE par la France, 451.

163. Note of D.H.A. Hannay, United Kingdom Delegation to the European Communities, November 24, 1967. PRO/FCO 55/47, TNA.

164. Council of the European Communities: Technology. Telegram No. 338 from Brussels to Foreign Office, November 1, 1967. PRO/FCO 55/59, TNA.

165. Piganiol, "Scientific Policy and the European Community," 354–365.

166. Albert H. Teich, "International Politics and International Science: A Study of Scientist's Attitudes." (PhD diss., Massachusetts Institute of Technology, 1969), 173.

167. Le Courrier du CNRS, 21 (1976), 3–8. The author's translation from French.

CHAPTER 4

Euratom: The Troubled Forerunner of Community Research Policy

The willingness of the European Community (EC) to enlarge its scope of action in research and technology that manifested itself in Luxemburg on October 31, 1967, involved a striking paradox: only a few weeks after that historical Council, research was subject of another ministerial meeting, characterized, however, by an entirely different atmosphere. On December 8, the Council gathered to discuss the future of the European Atomic Energy Community, Euratom. For the promoters of integration, the event and its outcome could hardly have been more disappointing: after lengthy and contentious discussions, an interim research program covering only the subsequent year and a draft research budget with cuts of 50 percent were approved.[1] Furthermore, the introduction of the principle of *volontariat* or *à la carte* cooperation enabling the member states to participate only in projects that they deemed to be most appropriate for their interests signified a remarkable diminution of the Community dimension in Euratom's activity. The dramatic contrast of this result to the earlier declarations on research policy was highlighted by *Frankfurter Allgemeine* which in the aftermath of the meeting concluded: "The discussion has made clear how long and laborious the road to a European research policy, which everyone is so urgently calling for, still is."[2] How can this discrepancy be explained? Why were the EC member states willing to proceed with the plans to expand the Community's activities in research while letting the main venue of common activity in the field decline?

© The Author(s) 2020
V. Mitzner, *European Union Research Policy*, Europe in Transition:
The NYU European Studies Series,
https://doi.org/10.1007/978-3-030-41395-8_4

The coincidence of the escalation of Euratom's difficulties in the mid-1960s and the enthusiasm for developing a general research policy was no accident. The disagreement in the nuclear sector not only constituted a major political problem that strained the relations between the six EC member states and threatened the image of the Community as a whole, but also endangered the existence of important Community institutions such as the Joint Research Centre (JRC). The urge to find new tasks and a new purpose for Euratom's institutions was an important driving force behind many proposals for a common research policy. Moreover, Euratom served as a crucial institutional and political basis for new activities. With Euratom, the Community did not need to start from scratch, but could build on precious scientific experience and expertise. This chapter discusses the intersection of the debate of a common research policy and the crisis of Euratom, and the way the political struggles in the nuclear sector paradoxically came to promote European integration.

Euratom was established together with the European Economic Community (EEC) by the Rome Treaties in March 1957. For its six original member countries (Belgium, France, Italy, Luxemburg, the Netherlands, and the Federal Republic of Germany), the creation of the two communities signified an attempt to overcome the political deadlock that had followed the failures of the military (European Defence Community, EDC) and political (European Political Community, EPC) initiatives for furthering European integration. From today's perspective it is interesting that initially, many placed their greatest hopes on Euratom rather than on the Common Market, which was perceived not only as awkward to realize but also as an affair of lesser importance. Rather than the EEC, it was Euratom that was believed to be the vehicle for relaunching the ideal of a gradual unification of the European continent.[3] The two Communities would, nevertheless, be mutually reinforcing, and they shared the goal of contributing to Europe's economic growth and political stability.[4]

The setting up of Euratom took place in a specific context of postwar enthusiasm for nuclear technology. Indeed, it was anticipated that nuclear science would give rise to a new technological and industrial revolution transforming the whole productive system. Furthermore, the increasing openness of nuclear politics manifested by President Eisenhower's "Atoms for Peace" speech[5] and by the two international conferences held in 1955 and 1958 in Geneva, reduced obstacles for international cooperation in the field.[6] An important factor determining the spirit of the times was also the estimates that there would be a radical increase in the Western European energy consumption and the consequent concerns about securing sufficient energy supply. In the aftermath of Suez crisis of 1956, resorting to nuclear technology, which would enable Europeans to increase their independence from external fuel sources, seemed as a particularly promising strategy to respond the perceived challenges.[7] But there was also an additional rationale behind Euratom's creation: for France, Euratom was a convenient means of gaining access to American resources and preventing Germany from going for it alone in a field of remarkable political and military importance. The Euratom treaty involved the setting up of a special agency with a monopoly of import and distribution of nuclear fuels, and a special inspection system. This matched the French desire not only to keep an eye on Germany, but also to gain greater European independence vis-à-vis the United States. A decisive factor behind the US support of the project,[8] the common control mechanism actually increased the freedom of the European countries that in their bilateral agreements with the Americans had been tied to strict procedures of surveillance.[9]

However, within only a few years after the signature of the Rome Treaties, the situation changed considerably. Already in 1959, Western Europe faced a coal surplus accompanied by the simultaneously falling prices of imported energy. Consequently, the competitiveness of nuclear power compared to more traditional power sources became more and more questionable.[10] Furthermore, it soon became clear that instead of focusing on the development of a coordinated Western European nuclear industry, the governments of the Six prioritized the development of their national nuclear programs. Most striking in this regard was France, where the government embarked on an ambitious civil and military nuclear technology program that left little room for supranational arrangements. Contrary to the views of Euratom's "founding fathers" of nuclear technology as a virgin territory providing a bountiful basis for European integration,[11] the field soon became drained by diverse national ambitions.[12] Euratom seemed to have been based on a fatal miscalculation.

Reconsideration of Euratom's fundamental goals thus appeared unavoidable. Consequently, the major aim of facilitating the growth and development of a European nuclear industry was gradually jettisoned, while research was given increasing emphasis. But common research activities necessitated reconciliation of different national approaches and objectives and therefore compromises and mutual sacrifices for which the general atmosphere in the EC was not very favorable. Disputes concerning *juste retour* (battles for returning to the member states funds equivalent to their financial contribution to the Community) also pervaded Euratom. Furthermore, the continuing disagreements over a number of technical questions, especially the choice of reactor type (the American light water or the Franco-British natural uranium—gas-graphite),[13] prevented the member states from committing to common long-term goals.

The difficulties stemming from diverging national interests were further aggravated in 1964 when the rise in costs and the changed economic and technological circumstances made it necessary to revise Euratom's second research program. The issue was crucial since Euratom's research activity was based on a multi-annual research agenda included in the Article 215 of the Euratom Treaty. The first program had a budget of 215 million Units of Account (UA)[14] and was divided into two parts: the initial phase concerned research in Euratom's JRC, while the second involved external contracts. The activities of the JRC were conducted in the four Europeanized national research centers, which were Ispra in Italy, Karlsruhe in Western Germany, Geel in Belgium, and Patten in the Netherlands. The external contracts, on the other hand, fell into three different categories: research contracts, which were standard contracts for a particular or generally limited project that the Community entrusted to an outside body; association contracts, which included participation in the management and operation of a number of large-scale research projects, typically already underway in a national context; and contracts of participation under which the Community was directly involved in the promotion of nuclear industry. The second five-year research program, adopted for the years 1963–1967, did not differ in any fundamental ways from the first program. The major difference was in the size of the budget that in the second program amounted to 425 million UA to which the unspent 20.5. million UA from the first program was added.[15]

The negotiations on the program revision proved long and contentious, and the resulting decision that was finally reached in June 1965, included only a very limited budget increase.[16] What is more, the program and budget revision planted the seeds of some crucial misunderstandings that were to seriously complicate the functioning of Euratom during the following years. With the embitterment caused by the first budget and program revision still fresh in mind, nobody seemed to be surprised when, only a year later, the emergence of a new budgetary deficit in the Community's fast reactor association program, the most important single research activity,[17] led to a new political impasse.[18] The Commission's proposals for a supplementary budget for the final part of the year 1966[19] and for the budget increase for the year 1967 required a new round of lengthy and bitter discussions among national representatives.[20] While in December 1966, the Council finally managed to concur on a supplementary budget,[21] the issue of budget increase for 1967 continued to divide the Community.[22] The agreement on the budget was generally seen to require a new revision of the second five-year research program, which not only has proven to be tricky but also necessitated unanimous decision.[23] Reaching a consensus was difficult in the political atmosphere characterized by mutual suspicion and distrust.

The prolonged discussions on the budget and the research program revision delayed serious planning for Euratom's future after the expiration of the second research program in 1967. Although in November 1967, the Commission finally proposed an interim program only covering the year 1968[24]—hoping for a new multi-annual program within the next six months[25]—it was obvious that the once so promising Atomic Energy Community was drifting around with no long-term agenda. This was an embarrassing reality that the European and American press made no effort to veil: whereas the German *Frankfurter Allgemeine* described the discussions in the December 1967 Euratom Council as long and fruitless and estimated there being little hope for reaching a better agreement in the course of the following spring,[26] *The Times* reported that the meeting highlighted the "decline of Euratom."[27] Even a more skeptical tone was adopted by *The Washington Post* that quoted the pessimistic words of one of the Council delegates: "If this [Euratom] were a private company instead an international organization, it would have collapsed tonight."[28]

Floating without a clear vision was perilous in a situation where an adjustment to the dramatically changed political and technological situation was calling for a rapid revision of the Community's goals. A recurrent statement in the discussions on Euratom's future was that nuclear technology had reached its industrial phase and would now be produced in a competitive environment. Research and development in the field would no longer belong to public institutions alone but would instead be more and more influenced and determined by (private or public) industrial enterprises unwilling to share their knowledge for the benefit of others.[29] This would have an important impact on the Community, which would increasingly find itself in competition with different national institutions. Therefore, many observers drew the conclusion that the goals and strategies of Euratom's research activity should be reexamined and the Community's role even reduced.[30] The third research program would be fundamentally different from the previous two.[31]

This necessity of a large-scale revision of Euratom's functions made the drafting of the third multi-annual program particularly challenging. What was at stake was nothing less than the whole mission of the Community and its place in the rapidly changing political, economic, and technological environment. Moreover, in a situation characterized by sharply differing national views and fading interest on the part of the member governments, the chances of discovering rapid solutions were rather vague. In France, the rise to power of General de Gaulle had transformed the previous, somewhat ambiguous but still constructive policy into outright attempts to dominate and control the Community. After 1964, France no longer respected Euratom's jurisdiction and concluded agreements for fuel supply without asking for the Community's assent.[32] In the late 1960s, the French subsequently blocked the negotiations on new programs by stubbornly opposing the various compromise proposals drafted either by the Commission or by one of the five governments. In 1969 Jean Rey, the president of the Commission, openly accused Paris of selfishness and warned about the consequences of a withdrawal from the common program in the pursuit of national interest. Rey also emphasized that Euratom's expenses constituted only 5 percent of the financial benefits France drew from the common agricultural policy.[33] How could France expect its partners to continue paying for exports of French wheat if the French government did not want to help them to develop nuclear capacity that they could not do on their own? These pleas had little effect: François-Xavier Ortoli, the French research minister, responded that Rey was exaggerating the problem, while repeating that Paris had little interest in Euratom's current program that hardly benefited France.[34] It is important to note that the deep-rooted hostility toward Euratom reached the highest echelons of France's political decision-making structures. The attempts of Prime Minister Michael Debré (1959–1962), for instance, to get rid of the Community altogether, have been well documented by historians.[35]

There were many reasons for the negative French attitude toward Euratom. The resurgence of French nationalism and the Gaullist distrust of supranational institutions were certainly the most prominent ones. But in nuclear research, France was also the undisputed leader in the continental Europe: already in 1956, the combined budget of its five partners represented only 25 percent of the budget of the French CEA (Commissariat à l'Energie Atomique).[36] Consequently, there was not very much France could expect to get back from the other Euratom member countries, all of them still being novices in the field. Moreover, pursuing their own ambitious nuclear weapons program, the French did not want to be tied by Euratom's inspection system. Even though Euratom's surveillance did not cover military research, Paris regarded supranational control as a violation of national sovereignty. So, a mechanism that originally had been useful in controlling Germany and increasing European room for maneuver in relation to the United States, now seemed to be working against France's interests. Very tellingly, France's refusal to renew the mandate of Etienne Hirsch, Euratom's enthusiastic second president, centered around the dispute on nuclear inspection.[37]

But also in Germany, especially in private industry that dominated the German nuclear programs and development, Euratom was often seen as an unwelcome instrument of state control and a symbol of the type of energy production for which, due to the national richness in other energy sources, there was no compelling need.[38] In fact, Germany had never been crazy about Euratom. Michael Eckert has shown how Bonn, instead of enthusiastically supporting initiatives for a supranational European framework, preferred from the outset bilateral arrangements with the United States. Only in the face of US pressure did the federal government agree to support the Euratom Treaty.[39]

The resurgent French and German national interests were most apparent in the plans for fast breeders, the vital third-generation reactors. Here both governments decided to develop their own separate fast breeder programs. The Commission managed to negotiate contracts of association in both cases, but it was left to cope with two different and rival efforts in which the European dimension had little room.[40] Similarly, the Community member states opted for cooperation outside the Atomic Energy Community in uranium enrichment. Throughout the 1960s, different plans had been developed to create a European installation for enriched uranium, which was the necessary fuel in European nuclear energy plants. Britain and France had small gaseous diffusion plants, but the output of these plants was devoted to military purposes, leaving the United States as the only Western Bloc country with production capacity for civilian use. European cooperation on uranium enrichment was thus intended to reduce the dependence on American deliveries. Euratom was active on the issue, and a plant was actually planned as a Franco-German venture within the Community framework. But these plans came to an end with Britain's irresistible offer to Germany. In November 1969, Britain, Germany, and the Netherlands reached the final agreement in tripartite talks on the construction of a gas centrifuge to produce enriched uranium.[41]

By the end of the 1960s, the increasing nuclear nationalism and the distrust in the Community had led to a situation where the most important nuclear research activity of the six member states took place outside Euratom's gradually shrinking research program and sometimes even with partners that were not part of the Community. This coincided with a steady exodus of the most talented researchers from Euratom's research facilities to private enterprises, which were able to provide better infrastructure and more secure working conditions.[42]

Crucially, the prospects for Euratom's future were influenced also by the simultaneous initiatives to create a common research policy in a more general sense. It was generally perceived that Euratom had to be situated in an extended, though still largely unspecified, framework. In the Council discussion of December 8, 1967, the Belgian delegation stressed the need to avoid any decision that would risk leading the Community's nuclear policy into a direction incompatible with the political goals agreed in the Luxemburg Council about a week earlier. According to the Belgians, the reform of the JRC's activities had to be determined by the choice of the primary Community objectives in the non-nuclear fields. The French Research Minister Schumann issued an almost identical statement, expressing his wish that the Council would not undermine the hopes raised in Luxemburg but rather confirm the political intentions expressed in that meeting.[43] Yet, serious political consideration on a common research policy had barely begun, and there was no clear view of where Euratom would sit in the new framework. In October 1967, Commissioner Hellwig argued that the future activities of Euratom should be defined in the perspective of the *eventual* new orientation of the Community's activities in the domain of research in general. At the same event, the German delegate Pretsch stated that the Community's activities in the nuclear sector could be complemented by other tasks *in case* the cooperation between the member states in the domain of scientific and technical research leads to a certain intensification of common activity.[44] Uncertainty continued to be the defining feature of the Community's discussions on a general research policy.

While this might have discouraged the member governments from making definite decisions on Euratom's future,[45] it is noteworthy, however, that in the discussions on the research program in 1967, the question of Euratom's new institutional position was almost entirely absent. Euratom was living hand to mouth, and consequently, the quest for tangible solutions was imperative. More urgent than trying to figure out where Euratom could sit in a larger and still very vague picture, was the necessity to prevent national disagreements from disturbing the Community's day-to-day functioning.[46]

But if the prospect of a general research policy was not the most prominent concern in the discussion on Euratom's new program, Euratom and its crisis became a useful instrument for the promoters of extended Community activity. The troubles of the atomic energy community provided a good argument for a structural overhaul of the Commission research administration and the setting up of institutional structures that could serve more ambitious political agenda. In the fusion of the three Commission Executives in July 1967—an institutional reform that merged the Commissions of the EEC and Euratom, and the High Authority of the European Coal and Steel Community—the Commission practically demolished Euratom to create a General Directorate of General Research.[47] A central argument for the DG General Research was that it would allow a more comprehensive view of the so-far scattered research activities. Euratom's limited framework was not ideal for responding to the new challenges, especially that of the technology gap, which necessitated a more comprehensive approach to science and technology.[48] The enthusiasm of the Commission in seizing this opportunity was openly revealed by the special councilor of the Commission President Hallstein, who in December 1966 declared that "[t]he Commission of the EEC is in favor of everything that can facilitate, in the context of the merger of the Executives and the merging of the Communities, the extension of research tasks."[49]

While the fusion of the Executives mostly took place as the Commission's internal affair,[50] some member governments also supported the idea of regrouping all bodies responsible for research and development within a single administrative framework.[51] For instance, the German Foreign Ministry saw it as a solution to the Community's problems in the nuclear field:

> We hope that the merger of the institutions, and even more so the fusion of the Communities, will resolve the problems Euratom faces. In particular, the economic policy side of nuclear energy development should be integrated into a common energy policy, and the Community's research activities should be extended from the slowly narrowing nuclear energy sector to other areas of scientific and technological research.[52]

The Ministry's message was very clear: the merger of the Community Executives should be used to end Euratom's distress and launch new activities in the field of research.

But the fusion of the Executives and the creation of the DG Research were not the only opportunity for general research policy stemming from Euratom's distress. In the mid-1960s, finding a new purpose for Euratom's joint research institutions became a compelling problem that pushed for creative solutions outside the nuclear fields. More generally, Euratom was thought to be used as a basis for general research policy.

Among the first presenting this idea was—surprisingly perhaps— Euratom's skeptical President Pierre Chatenet, who in an interview published in the French newspaper *Le Figaro* in April 1966 stated: "The disappointing, seemingly fruitless but rich experience of Euratom is coming to an end. Do not throw the baby with the bath water."[53] While expressing disillusionment and skepticism with the future of Euratom, Chatenet also highlighted new prospects for action.[54] This was met with significant enthusiasm in the European Parliament,[55] which published a resolution stating that central role should be assigned to Euratom in the management of Community's research projects as well as in the coordination of common programs. Another resolution called for an extension of the competences of Euratom's JRC to cover all sectors of research.[56] On December 8, 1967, the Council decided to enlarge the JRC's activities to non-nuclear fields.[57]

This was clearly motivated by the distress of the institution. The insecurity concerning Euratom's next research program and the budget cuts affected the center in a very direct way. In August 1966, Jules Guèron, General Director at Euratom, wrote:

> The Joint Center for Nuclear Research (JRC) as it exists at the end of the second plan is therefore a tool that as such can no longer find a place in a third [Euratom research] program. Since the removal of JRC establishments is impossible, for reasons related to social order and regional economy, the only available solution is to reconvert the centers. This could be carried out according to the following principles: the expansion of the centers would be stopped until further notice... a part of the centers would be converted to other non-nuclear Community activities and funded through other budgets... the centers would gradually abandon research in the field of reactors to become more and more Community facilities available to the Member States or industries.[58]

With Euratom's third research program, the JRC would hit the wall. The center either was reformed and repurposed or had no raison d'être.

Neither the creation of the DG Research nor the Council decision on JRC's reform immediately resolved the problem of Euratom's research centers. During the next two years, the problem of finding jobs for those affected by the cuts in Euratom's research program continued to concern the Commission and especially the Italian government. The Italians were particularly worried about the future of Ispra, the biggest Community joint research center that in 1969 employed 1590 persons out of the total JRC staff of 2700 and was situated in Northern Italy. There was also increasing public pressure to find a solution: the decision of the Council in March 1969 to reform the administration by the end of that year provoked public protests, among them a hunger strike of ten workers of Ispra. This was too much for the representative of the Italian government who in July appealed to the other EC governments to accelerate the works that would lead to new joint engagements both in nuclear and non-nuclear research. For the support of his case, he resorted to the argument of European retard: a bold common activity within Euratom would eventually fill the transatlantic technology gap.[59]

The Italian pleas echoed the written appeal of the JRC's workers, sent to Commission's President Jean Rey in May 1969. According to that document, the decision to cut the center's personnel, "even before a scientific policy is defined at Community level," could only have the consequence of:

> the start of the dismantling of this European body; the ruin of ten years of concrete cooperation and, consequently, the prospects for a Community research policy; the disappointment of the young generation that is forming in this environment and in the European School associated with it; a crisis of confidence vis-à-vis the Community institutions and their bodies.[60]

Both the Italians and the spokesmen of the JRC explicitly connected the survival of the center to the prospect of developing general research policy for the Community. The Italian interest in the issue stemmed from the continued distress of the country's research. Between 1970 and 1975, Italy was the Community member state with the lowest increase in investments for research. A revealing OECD report—which for the Italian government was so embarrassing that it first tried to block its release—showed that the Italian expenditure per head for research and development was $10.7 in 1967, whereas the equivalent figure for 1965 in France was $37.9

and in Britain $39.8. The report also disclosed that in the late 1960s, only 0.6 percent of Italian GNP went to research, compared with 2.0 percent in Britain and 1.4 percent in France. The number of qualified researchers and technicians per thousand of population was equally low in Italy, where the figure was 6, whereas the figures for Britain, Sweden, and Germany were 29.4, 21.6, and 18, respectively. "The report makes it perfectly clear that Italy is most delinquent among Europe's industrialized nations when it comes to supporting research and education," D.S. Greenberg wrote in *Science* magazine in 1969. "Over the past 2 years there has been little if any improvement, and in many respects the plight of research and education in Italy has deteriorated as a result of strikes, administrative chaos, and a general deterioration of public administration."[61]

Widening the scope of the Community's research activities on the basis of Euratom was nevertheless problematic. It was unclear whether the Euratom Treaty could be used for the new non-nuclear activities. The Commission rebuffed these doubts. In the meeting of the Political Commission of the European Parliament in December 1966, Commissioner Robert Marguelis said: "The Euratom Commission is, however, of the opinion that an extension of Euratom activities to other scientific or technical fields is possible without modification of the provisions of the Euratom Treaty."[62] A similar, though slightly more careful, statement was made the by the vice-president of the Euratom Commission Antonio Carrelli only a couple of weeks later: "A certain extension of the competences listed in the Treaty can be envisaged."[63] In fact, the Commission's argument was rather that the Euratom Treaty offered a basis on which future research politics could be built. This view was also propagated in the European Parliament, where Eduardo Battaglia declared: "The language of the Euratom Treaty contains, in its extensive interpretation, the basic elements necessary for the exercise of a Community activity in all parts of the scientific and technical collaboration."[64] Yet, the French and Dutch governments saw the issue differently.[65] Paris in particular did not hesitate to employ the juridical argument as a pretext to block reforms that would have given the Community more scope.[66]

Indeed, despite the support of Italy, the Commission, and a few other member states,[67] the atomic energy community's existence continued to be troublesome. In December 1969, an internal Commission document evaluated the situation: besides "certain positive aspects," such as the prolongation of the common research program by one or two years, the decision to refrain from staff cutbacks during that period, the agreement to extend the JRC's works to sectors other than nuclear energy and finally, the launching of a study to improve the management of JRC, the improvements were rather cosmetic:

> the patient who was the Euratom program leaves the clinic after having an operation of opening and noting the gravity of the case, and then to closing again and saying "we will study which remedy will be applied within one or two years"– without know which one. I fear this does not indicate a quick cure.[68]

And this dim prediction proved quite right: despite the Commission's subsequent efforts to give the Community a new start, in December 1971, the Council failed again to agree on a new multi-year research program for Euratom. Speaking to a press conference after the Council meeting, Commissioner Spinelli lamented that the ministers had shown no interest in a thorough discussion on the program as a whole.[69] As before, the sole concern seemed to be arguing about minor points, while France agreed to participate in only half of the program.[70] The wrangle led to one more one-year arrangement with not much purpose other than keeping the atomic energy community alive until the agreement on its long-term program could be reached.[71] This had the immediate effect of strengthening the general feeling of uncertainty and disorientation that prevailed among the employees of Euratom's four joint research centers. The public protests also continued in these establishments in the following year.[72]

Agreement on a new multi-annual research program was finally reached in February 1973, after fourteen hours of tortuous deliberations that lasted until dawn. The new program of $157 million[73] covered four years and provided work for 1649 scientists in the Community's joint research centers. This guaranteed that no severe staff cutbacks in the JRC would take place: only fifty scientists were expected to be dismissed. However, the outcome was not exactly what Paris wanted. France, which still regarded Euratom as a useless obstruction standing in the way of its national nuclear ambitions, had been advocating, together with Britain and the Netherlands, considerable pruning of the staff and operating costs. This pushed the Community only further to develop new activities in the field of research. Following the meeting, *The Times*, noted: "Nuclear research is a hungry business. It was clear from a joint British and French objection last week to giving Euratom a large budget that they shared this view. The long-term future for the agency must hang on the ability to build expertise in other fields."[74] This was the widely shared conclusion in Brussels as well.

Some historians see Euratom's difficulties mainly as a constraint on the development of a wider conception of common research policy. Certain member countries, they argue, first wanted to settle the problem of nuclear research before making decisions on a wider political framework.[75] In part this is true: occasionally, the situation in the nuclear field intruded on the discussions on a general research policy in a manner different from what this chapter has shown. During the meeting of the Community's research ministers in Luxemburg in 1967, for instance, the German delegate Hans von Heppe said that stretching the cooperation into new domains was conditioned by the political will of the Community's member states to overcome the difficulties faced in nuclear research.[76] The Dutch Ambassador Spierenburg made a similar argument in the Council of June 5 and 6, 1967. He found it illusory that rapid progress in the question of scientific cooperation would be possible as long as the problems of Euratom's budget and the second research program remained unsolved. Spierenburg even indicated that providing that Euratom's crisis continued, there would be no guarantee of collaboration on the Dutch side in the examination of the issues related to research.[77] Even the resolution of the Luxembourg Council 1967 stated that "on the occasion of these deliberations on scientific and technical research, the Council reaffirms the importance it attaches to ensuring that constructive decisions are made rapidly on Euratom's future research activities."[78]

Although Euratom's troubles were pictured as an obstacle to the attempts to develop general research policy, it is difficult to prove that these statements went beyond political rhetoric. Indeed, the causal connection seems to be the other way around: there are a number of factors, which indicate that the hardships in the nuclear field eased the move toward a more comprehensive research policy. The administrative reform of the fusion opened a window of opportunity to set up a new DG with a broader mandate. Moreover, the problem of finding jobs for a number of persons, who were threatened to be laid off in Euratom's research centers, and the urge to find a new purpose for these institutions, led the Commission and some member governments to consider new functions for the JRC. Euratom's difficulties actually facilitated the project of a general research policy.

The importance of Euratom's legacy for the Community's later development has also been shown by other researchers, such as Andrew Barry and William Walters, who stress the functionalist heritage of that Community institution.[79] But even closer to this analysis of Euratom's immediate impact on the launching of a general research policy comes Christian Fischer-Dieskau: "There is no doubt that... the European Community research activity developed under the Euratom Treaty is the basis of our current research policy."[80] But it was more than the mere existence of the Atomic Energy Community that enabled the EC to embark on its new adventure on research policy. Euratom's problems came as a decisive push factor that led politicians to look outside the box to find creative solutions. For the promoters of integration, Euratom's crisis offered an excellent chance to realize their vision of an activity that had no place in the founding treaties. Rather than being obstacle to initiatives, it was a unique opportunity to begin something genuinely new.

NOTES

1. Commission of the European Communities, *First General Report on the Activities of the Communities 1967* (Brussels-Luxembourg: February, 1968), 298.
2. *Fankfurter Allgemeine*, December 9, 1967. The author's translation from German.
3. Luca Guzzetti, *A Brief History of European Union Research Policy* (Luxembourg: European Communities, 1995): 7.
4. John Krige, "The Peaceful Atom as Political Weapon: Euratom as an Instrument of U.S. Foreign Policy in the 1950s." *Historical Studies in the Natural Sciences*, 23 (2008): 3.
5. Guzzetti, *A Brief History*, 8.
6. Maurice Vaïsse, "La coopération nucléaire en Europe (1955–1958). Etat de l'historiographie," in *L'énergie nucléaire en Europe. Des origines à Euratom* eds. Michel Dumoulin, Pierre Guillen, and Maurice Vaïsse (Louvain-la-Neuve: Euroclio, 1991): 99.
7. Guzzetti, *A Brief History*, 10–11.
8. Christopher Layton, *European Advanced Technology. A Programme for Integration* (London: George Allen & Unwin, 1969): 105, 107; Henry R. Nau, *National Politics and International Technology, Nuclear Reactor Development in Western Europe* (Baltimore and London: The John Hopkins University Press, 1974): 96.
9. Krige, "The Peaceful Atom," 5–44.

10. Lawrence Scheinman, "Euratom: Nuclear Integration in Europe," *International Conciliation*, 563 (1967): 27.
11. Guzzetti, *A Brief History*, 8.
12. Nau, *National Politics*, 97.
13. The French wanted the Community to adopt its reactor design, whereas the five other member countries opted for the American model. The Commission, for its part, tried to keep a neutral and equal position, which, however, rather created uncertainty and indignation than facilitated the choice. Guzzetti, *A Brief History*, 29.
14. The Community Unit of Account served as a standard monetary unit of measurement before the introduction of the Euro in 1999. Until the end of the fixed exchange rates in 1971, the Community Unit of Account was worth of 0.88 grams of gold, i.e. one US dollar. Guzzetti, *A Brief History*, 32.
15. Ibid., 16–18; Scheinman, "Euratom," 14.
16. From 425 to 430.5 million UA. Ibid., 43.
17. *The Times*, June 5, 1967.
18. Euratom: Difficultés de programme et de budget. Memorandum de G. Guazzugli Marini, Strasbourg, le 21 juin 1967. Emilé Noël (hereafter EN)-000443, HAEU.
19. Euratom Council note: Avant-projet de budget supplémentaire de recherches et d'investissement de la Communauté pour l'exercice 1966 (crédits de paiement). BAC 26/1969-825, HAEU.
20. The budget revision for 1966 was first opposed by the German, Italian, and Dutch delegations that saw no justification for such a demand and argued that supplementary funds could be considered only in the framework of the budget for the following year. Projet de procès-verbal de la réunion restreinte tenue à l'occasion de la 116ème session du Conseil de la Communauté Européenne de l'Energie Atomique, Bruxelles, les mercredi 26 et jeudi 27 octobre 1966. BAC 118/1986-605, HAEU.
21. The decision was passed by the qualified majority vote of the Italian delegation still opposing. Procès-verbal de la réunion restreinte tenue à l'occasion de la 118ème session du Conseil de la Communauté Européenne de l'Energie Atomique, Bruxelles, les mardi 6 et mercredi 7 décembre 1966. CM2 1966-0081, HAEU.
22. Commission of the European Communities, *First General Report*, 296.
23. Euratom: Difficultés de programme et de budget. Memorandum de G. Guazzugli Marini, Strasbourg, le 2 juin 1967. EN-000443, HAEU.
24. Commission of the European Communities, *First General Report*, 298.
25. Extrait du projet de procès-verbal de la 13è réunion du Comité Consultatif de la Recherche Nucléaire du le 26 octobre 1967. BAC 180/1980-144, HAEU.
26. *Frankfurter Allgemeine*, December 9, 1967.

27. *The Times,* December 9, 1967.
28. *The Washington Post,* December 9, 1967.
29. This factor was seen to be at the core of Euratom's problems, for instance, by Commissioner de Groote in his address to the European Parliament on October 18, 1966. *Agence Europe,* October 18, 1966.
30. Remarques au sujet de l'établissement d'un IIIème programme pour Euratom. Bruxelles, le 29 août 1966. JG-108, HAEU.
31. *Agence Europe,* October 18, 1966.
32. Nau, *National Politics,* 100–101.
33. Procès-verbal de la réunion restreinte tenue à l'occasion de la 72ème session du Conseil, Luxembourg, le lundi 30 juin 1969. CM2 1969-32, HAEU.
34. Procès-verbal de la réunion restreinte tenue à l'occasion de la 84ème session du Conseil, Luxembourg, le mardi 28 octobre 1969. CM2 1969-48, HAEU.
35. Robert Frank, "Michel Debré et l'Europe," In *Michel Debré, Premier Ministre (1959–1962),* eds. Serge Bernstein, Pierre Milza, and Jean-François Sirinelli (Paris: Presses Universitaires de France, 2005), 308–309.
36. Krige, "The Peaceful Atom," 19.
37. Guzzetti, *A Brief History,* 24–25.
38. Scheinman, "Euratom," 33–37. Aufzeichnung: Berücksichtigung außenpolitischer Geschichtspunkte durch das Bundesministerium für Wissenschaftlichen Forschung (Beteiligung der Auswärtigen Amts), Bonn, den 20. Januar 1967. B 35-78, PA AA.
39. Michael Eckert, "Kernenergie und Westintegration. Die Zähmung des westdeutschen Nuklearnationalismus," In *Vom Marshallplan zur EWG. Die Einigung der Bundesrepublik Deutschland in die Westliche Welt,* eds. Ludolf Herbst, Werner Bührer, and Hanno Sowade, (München: Oldenburg, 1990): 313–334.
40. Layton, *European Advanced Technology,* 108–109.
41. Susanna Schrafstetter and Stephen Twigge, "Spinning into Europe: Britain, West Germany and the Netherlands – Uranium Enrichment and the Development of the Gas Centrifuge 1964–1970," *Contemporary European History* 2 (2002): 256–257, 260–262.
42. Procès-verbal de la réunion restreinte tenue à l'occasion de la 84ème session du Conseil, Luxembourg, mardi le 28 octobre 1969. CM2 1969-48, HAEU.
43. Projet de procès-verbal de la 15ème session du Conseil tenue à Bruxelles, vendredi le 8 décembre 1967. JG-111, HAEU.
44. Projet de procès-verbal de la treizième réunion du Comité Consultatif de la Recherche Nucléaire tenue à Bruxelles le 26 octobre 1967. BAC 118/1986-2965, HAEU.

45. Some scholars argue that the development of a common research policy complicated and delayed the negotiations on Euratom's multi-annual research program. Guzzetti, *A Brief History*, 29.

46. Ibid.

47. Heinrich von Moltke, interview by Julie Cailleau and Arthe van Laer, Tervuren, January 22, 2002. European Commission Memories, European Oral History, HAEU.

48. Parlement européen, Commission politique (W.J. Schuijt): Avant-projet de rapport sur le rôle de la Communauté européenne de l'Energie atomique dans l'Exécutif unifié, le 17 janvier 1967. CEAB2-2622, HAEU. Euratom was also connected to the technology gap debate in a more direct way, namely, as an eventual site or cause of the gap; the difficulties of Euratom were thought to weaken Western European performance in the field of nuclear technology, a sector of fundamental importance in the technological race. Parlement Européen, Débats. Compté rendu in extenso des séances n°87, session 1966–1967. Séances du 17 au 21 octobre 1966, 23, Parlement Européen, Débats. Compte rendu in extenso des séances n°97, session 1967–1968. Séance du 8 janvier 1968, 15–16.

49. Parlement Européen, Commission de la recherche et de la culture. Procès-verbal de la réunion du 19 décembre 1966. CEAB2-2621, HAEU. The author's translation from French.

50. The reorganization of the Commission took place as an internal affair of the Single Executive. Here Jean Rey, the new president of the Commission, who conducted personal discussions with the individual Commissioners about their future responsibilities, played a key role. Rey got the necessary approval of all the Commission members for the new arrangement, and so there was no general discussion on the Commission's competences in the different domains. The DG General Research was created without any significant political debate. Fritz Hellwig, interview by Veera Mitzner, Bonn Bad Godesberg, February 3, 2010.

51. Europäische Zusammenarbeit auf dem Gebiet vom Forschung und Entwicklung, Bad Godesberg, den 14. Februar 1967. B 138/3944, BAK.

52. Deutsch-französische Konsultationen auf Direktorsebene am 10. Januar 1967 in Bonn: Zukunft Euratom's, Bonn, den 5. Januar 1967. B 20-200 1167, PA AA. The author's translation from German.

53. *Le Figaro,* April 28, 1966. The author's translation from French.

54. Extrait du quotidien "Il giornale d'Italia" les 24–25 mai 1966. CEAB2-2756/3, HAEU; Parlement Européen, Débats. Compté rendu in extenso des séances n°86, session 1966–1967. Séances du 27 juin au 1er juillet 1966, 156–157.

55. Parlement Européen, Débats. Compté rendu in extenso des séances n°86, session 1966–1967. Séances du 27 juin au 1er juillet 1966, 169.

56. Parlement Européen, Débats. Compté rendu in extenso des séances n°87, session 1966–1967. Séances du 17 au 21 octobre 1966, 57, 81.

57. Résolution du Conseil concernant les activités futures d'Euratom. Annexe II au doc. R/1896/67, CM2 1967-825, HAEU.

58. Remarques au sujet de l'établissement d'un IIIème programme pour Euratom, Bruxelles, le 29 août 1966. JG-108, HAEU. The author's translation from French.

59. Procès-verbal de la réunion restreinte tenue à l'occasion de la 84ème session du Conseil, Luxembourg, le mardi 28 octobre 1969. CM2 1969-48, HAEU.

60. Du Comité de défense du CCR (Ispra) à M. Rey, Ispra, le 23 mai 1969. JG-105, HAEU. The author's translation from French.

61. D.S. Greenberg, "Italy: OECD Report Finally Emerges," *Science*, October 31, 1969: 587.

62. Procès-verbal, Commission politique, PE le 5 décembre 1966. CEAB2–2621/4, HAEU. The author's translation from French.

63. Parlement Européen, Commission de la recherche et de la culture. Procès-verbal de la réunion du 19 décembre 1966. CEAB2-2621, HAEU.

64. Parlement Européen, Commission de la recherche et de la culture (Eduardo Battaglia): Avis l'intention dans la commission politique sur l'activité de la Communauté européenne de l'énergie atomique dans l'Exécutive unique CEAB2-2622, HAEU. The author's translation from French.

65. Procès-verbal de la réunion restreinte tenue à l'occasion de la 92ème session du Conseil, Bruxelles, le samedi 6 décembre 1969. CM2 1969-786, HAEU.

66. Procès-verbal de la réunion restreinte tenue à l'occasion de la 72ème session du Conseil, Luxembourg, le lundi 30 juin 1969. CM2 1969-32, HAEU.

67. Germany eventually joined the demands for expanding Euratom's next research program to non-nuclear activities. Herrn Vizepräsident Dr. F. Hellwig: Vorschläge von Herrn Minister de Block – Zukünftige Mehrjahresforschungsprogramm, Brüssel, den 23. Oktober 1969. I-083 Hellwig 169/1, ACDP.

68. Note aux directeurs d'établissement et chefs de projet. Décision de programme du conseil de ministres en date du 6.12.69. (E.H. Hubert), Bruxelles, le 8 décembre 1969. JG-181, HAEU. The author's translation from French.

69. *The Times*, December 13, 1971.

70. *The Times*, December 8, 1971.

71. *The Times*, December 21, 1971 and December 22, 1971.

72. *The Times*, June 26, 1972.

73. Nau, *National Politics*, 121.

74. *The Times*, February 7, 1973.

75. Éric Bussière and Arthe van Laer, "Research and Technology or the 'Six National Guardians' for 'the Commission, the Eternal Minor,'" in *The European Commission 1958–72. History and Memories of an Institution*, eds. Michel Dumoulin et al. (Luxemburg: European Commission, 2007): 498; van Laer, "Vers une politique," 44.

76. Procès-verbal de la réunion restreinte tenue à l'occasion de la 11ème session du Conseil, Luxemburg, le mardi 31 octobre 1967. CM2 1967-73, HAEU.

77. Projet de procès-verbal de la réunion restreinte tenue à l'occasion de la 126ème session du Conseil de la Communauté européenne de l'énergie atomique, Bruxelles, les lundi 5 et mardi 6 juin 1967. BAC 118/1986-613, HAEU. The same argument was presented again by the Dutch in the COREPER meeting of 9 June 1967. Comité ad hoc du Groupe Inter-exécutif Recherche: Examen des résultats de la réunion du Comité des Représentants permanents du 9 Juin 1967. BAC 118/1986-1399, HAEU.

78. Problèmes de la recherche scientifique et technique dans les Communautés. Annexe IV au doc. R/1893/67. CM2, 1967-73, HAEU. The author's translation from French.

79. Andrew Barry and William Walters, "From EURATOM to 'Complex Systems': Technology and European Government." *Alternatives: Global, Local, Political*, 28 (2003): 311–312.

80. Christian Fischer-Dieskau, "Ziele und Methoden europäischen Forschungspolitik." In *Öffentliche Aufträge und Forschungspolitik*, eds. Karl Matthias Messen (Baden-Baden: Nomos, 1979), 35. Quoted in Michael Felder, *Forschungs- und Technologiepolitik zwischen Inter-nationalisierung und Regionalisierung* (Studien der Forschungsgruppe Europäische Gemeinschaften, Institut für Politikwissenschaft, Philipps-Universität Marburg, 1992), 83. The author's translation from German.

PART II

CHAPTER 5

The Years of Questioning

Economic growth constituted the initial driving force of postwar European science policy. Similarly, within the European Community, the first debates on the matter were firmly based on the objective of continuous economic expansion and the strong belief in the benefit of scientific activity for the pursuit of growth. In the late 1960s, however, the discursive context changed. Doubts emerged about the possibility and desirability of growth as well as the direct economic impact of scientific research. This opened a window for a profound reassessment of the fundamental goals and methods of science policy. In addition, with the rise of the environmental movement, science became associated with the environmental deterioration caused by large-scale application of technology, while new economic constraints imposed different political priorities.

This said, the assault on science per se remained weak. Rather than the authority of science, the new criticism focused on the ways in which science was used and questioned certain orientations of science policy. This pressure led to some important revisions of governmental activities in the field, but changes to the general political approach remained cosmetic. The political frame presenting research as an indispensable engine for growth—a major political goal of all industrialized countries—remained firmly in place throughout the 1970s. Fighting inflation, at least in some countries, momentarily challenged fast and continuous expansion as the primary objective of national economic policy,[1] and the new environmental and social concerns provoked more nuanced perspectives stressing

© The Author(s) 2020
V. Mitzner, *European Union Research Policy*, Europe in Transition:
The NYU European Studies Series,
https://doi.org/10.1007/978-3-030-41395-8_5

qualitative aspects of economic activity, but nothing seriously undermined the continuing appeal of growth policies of which research continued to be seen as the unfailing servant.

The second part of this book analyzes the struggle of the advocates for wider European Community research competences against the emerging challenges of science policy. Although some important changes in European politics seemed to create a fresh momentum for expanding the EC's activity into new domains, research remained the hostage of political disagreements in the absence of a shared vision for the future. Importantly, this debate took place in a context where the principles on which the blueprints of research policy had been based became increasingly difficult to defend. Caught in the murky space between continuity and change, the discussion dragged on drawing its strength from the rapidly changing global economic and technological conditions and the still insatiable political thirst for growth.

In the Western industrialized countries, the 1970s was an epoch of turmoil and transformation that marked a decisive break with the relative stable social, political, economic, and cultural framework of the postwar years. In many accounts, the period has been interpreted as one of crisis. For instance, Charles Maier, defining "crisis" as a situation in which "institutional arrangements of a society no longer deliver the results expected of them…and where normal corrective actions seem only to make the situation worse," presents the turmoil of the decade as a "crisis of industrial society" and the third systemic crisis of the twentieth century. According to Maier, in the decade and a half from the late 1960s to the beginning of the 1980s, Western industrial societies were caught in turbulence, in which the political economies developed to revive capitalism after the Great Depression, such as the Keynesian programs of government spending and the postwar welfare state, seemed to fail.[2] This created a sense of profound interconnection and erosion of confidence, which, again, led to a fundamental reconsideration of the economic and political axioms that had been regarded as self-evident since World War II.[3]

An important assumption that the "crisis" of the 1970s challenged was that of the possibility and universal desirability of continuous economic growth. During the 1950s and 1960s, the real growth of GDP had become one of the most important political and organization factors with global significance. In industrial societies, both in the East and in the West, by the end of the 1960s, economic growth was not only identified with the increasing wealth and standard of living but it also was regarded as "natural" and "healthy."[4] Growth stood at the heart of postwar materialism, the highly attractive and pervasive ideology of "more." As Harold James has noted, only "a few of this generation realized that this rather simple materialism was inherently unsatisfactory."[5] The mythical triangle of wealth, happiness, and peace was commonly taken for granted.

Toward the end of the 1960s, however, the popular ideology of "more" started to erode. The first indication of change in the growth debate came with the events of 1968. Besides the emergence of the global youth movement that shook traditional institutions and value structures across the world, the United States that year witnessed the most serious economic crisis since the Great Depression.[6] A result of a bundle of intertwined economic problems, such as the worsening balance of payments deficit and the inflation caused by the Vietnam War, the crisis revealed and contributed to the expiration of the American postwar economic hegemony and marked an end to American growth liberalism—the interpretation of liberal politics and growth economics that had been a defining feature of the optimistic 1960s.[7] This went hand in hand with the worrying observation that the economic theories followed by the Western governments appeared unable to respond to the new challenges.[8] The OECD experts described the new condition as "the precursor of saturation" and predicted that "a declining rate of growth is foreseeable within the lifetime of people now alive."[9] All of a sudden, it seemed as if the postwar period of unparalleled economic expansion had, in the end, been exceptional. The growth, after all, had its limits.

But it was not only the economies' ability to grow that provoked the new discussion of growth's limits. There was also the uncomfortable realization that continuous economic expansion had become ecologically unsustainable. Indeed, the late 1960s witnessed the explosive arrival of the environmental movement. Sparked off by the intellectual debate around some influential publications, such as Rachel Carson's antipesticide exposé, *Silent Spring* (1962), and accentuated by a series of ecological disasters in the late 1960s that powerfully exposed the fragility of the ecosystem, the new environmentalists became an important force challenging the postwar ideology of growth.[10] For example, in France, as Michael Bess notes in his study of French environmentalism, "at the beginning of the decade, the notion of safeguarding an endangered environment was still confined to the margins of society; by 1969, 'la protection de la nature' had become a familiar phase not only in the French mass media, but in political circles as well."[11] The change was similar in Germany, where the popularity of the new American environmentalist literature rapidly shifted the public attention from local environmental problems to global threats with no geographic or temporary boundaries.[12] In the United States, the Earth Day on April 22, 1970, mobilized millions of Americans across the country, effectively demonstrating the new awareness and attraction of environmental issues. During the 1970s, the environmental groups expanded and consolidated their efforts, which in many countries led to legislative initiatives, regulations, and legal precedents as well as to the creation of new institutions, such as the U.S. Environmental Protection Agency (1970) or the UK Royal Commission on Environmental Pollution and the Department of the Environment (1970).[13] Furthermore, the United Nation's 1972 Conference on the Human Environment initiated a multilateral political exchange on environmental issues and led to the creation of the United Nations Environmental Program, marking an important shift toward coordinated global action to protect the environment.[14]

The most authoritative attack on the growth of the postwar period[15] came from a team of researchers at the Massachusetts Institute of Technology (MIT). Sponsored by the Club of Rome—an esteemed group of scientists, business executives, educators, and technocrats from twenty-five nations—the MIT team produced in 1972 a study *The Limits to Growth*.[16] It issued pessimistic results: real-world resource and pollution problems could be kept manageable only if population growth and capital investment/output were quickly stabilized and held steady. The choice, if there was one, was between self-imposed limitation on growth or disaster. In a very short period of time, the report became an international bestseller; by the end of the decade, four million copies were in print in thirty languages.[17] With its highly publicized launch and its eminent institutional backing, the message of the MIT group generated a wide resonance with the public and made a crucial contribution to the changing of the political discourse on economy, environment, and science.

What makes the Club of Rome report particularly interesting is its OECD origins. The organization whose identity and raison d'être heavily rested on the promise of economic growth, now provided the institutional framework for international intellectuals challenging the rationale of endless economic expansion. Matthias Schmelzer has shown how a group of scientists and bureaucrats around the OECD science branch and its Director-General Alexander King proved instrumental in launching the Club of Rome. This group:

formed a shared set of normative and principled beliefs that revolved around the interconnected crisis-phenomena of the problématique, the limits to exponential quantitative growth, and the need for long-term planetary management; they generated shared causal beliefs, in particular regarding the detrimental social and ecological effects of uncontrolled growth, technology, and markets; and they pushed a common policy enterprise both within the OECD and in other organizations and contexts, in particular the Club of Rome.[18]

Yet the non-conformist discussions of King's circles were not welcomed with open arms. In particular, the top macro-economists of the Economics Department dismissed and opposed the efforts to redefine the organization's economic policy goals. Moreover, the OECD never abandoned its growth objective, which it came to consider compatible with environmental protection and a necessary precondition for progress. In the end, the critical perspective on the quantitative growth paradigm became marginalized within the organization and remained without significant effect.[19]

There was similar continuity in national discussions. Especially in the United States, a country where materialism and economic success were not only ideational fashion items of the postwar era but also a crucial part of the national identity and cultural and historical self-understanding. Toward the end of his presidency, Richard Nixon, concerned about what he recognized as negativism, loss of self-confidence, and estrangement from traditional values and the demise of patriotism, embarked on an ambitious crusade for growth. For Nixon, it was growth that would cure the nation of its mental crisis and guarantee the United States continued leadership in the "new world economy," increasingly characterized by fierce global competition.[20] This was despite the increasing skepticism of the social benefits of growth: data of public opinion surveys showed little relationship between average per capita income and degree of happiness. Americans, in fact, believed themselves to be less happy in 1970 than they were in 1957—although they were considerable richer.[21]

Growth remained on political agendas also on the other side of the Atlantic Ocean. As opposed to scientists and other intellectuals who expressed concern about the consequences of economic expansion, most political decision-makers in Europe did not seriously discuss the problem of "limits" or drastically revise their approach to growth.[22] In Germany, where the debate had a relatively late start,[23] the adherence to growth remained remarkably strong and almost unaltered. The growth paradigm, which the Grand Coalition of 1966–1969 forcefully embraced, was also systematically supported by the successive governments. This is of course not to say that the German political decision-makers would have been totally unresponsive to the concerns surrounding growth. As a consequence of the alarm sounded on the "limits" of growth, there was an increasing emphasis on the qualitative aspects of growth in the political discourse.[24] But as Richard Steurer notes, rather than a paradigm change, this strikingly uncontroversial discursive move only represented a corrective measure, a delicate shift in consensus within the still solid "growth coalition." There was no real growth controversy in the German politics. "Economic growth," Steurer concludes, "was still at the beginning of the 1969–1982 social-liberal era a paramount political goal."[25] Indeed, for the new German administration of 1974, environmental concerns appeared as an unwelcome break on growth, and Chancellor Helmut Smith even lamented on an environmental policy that was getting out of hand.[26]

In the Netherlands, the discussion on growth's limits started relatively early, and it not only reached a remarkably wide audience but also provoked a noticeable reaction in the major political parties. Steurer explains this by the media attention given to the work of the MIT researchers already early on. The reportage *The Limits for Our Globe*, that the Dutch TV broadcasted in November 1971, attracted seven million viewers, which was about half of the population. As a consequence, in the Netherlands, the Club of Rome report sold better than in any other country. Moreover, the growth debate in the Netherlands involved famous political personalities, such as the former Dutch Minister of Agriculture and the European Commissioner Sicco Mansholt, who became a visible advocate for the report's message and attracted popular attention to the report. In March 1972, Mansholt was appointed the president of the European Commission.[27] That said, also in the Netherlands, the debate on growth showed remarkable continuity. The most striking effect of the critical voices, mainly expressed by the new leftist parties, was the addition of "quality of life" to the national economic objectives. Just as in Germany after 1974, the political atmosphere in the Netherlands soon adopted more apprehensive attitudes toward growth.[28]

The political attention to the crisis of growth thus proved short and surprisingly superficial. However dramatic and painful the awakening to the world of "limits" had been, people and governments still wanted more growth and refused to seriously consider tradeoffs between economic and socio-environmental objectives. This effectively curbed the impact of groups calling for a revision of the growth paradigm. As Benjamin Kline notes in his history of American environmental movement, "the progress of the environmental movement remained subject to economic concerns – a dominant theme of American life and politics since the colonial period." Indeed, by the end of the decade, the movement had lost its momentum. The energy crisis of the mid-1970s and the new political initiatives of President Ronald Reagan "demonstrated that environmental issues, though important, were subordinate in the public mind to material living standards and economic security."[29] In America, the political window for rethinking growth had closed.

As shocking as the conclusions of the Club of Rome report were, they became easily ignored or absorbed by more immediate political concerns. This happened also in the European Community. Although a charismatic Commission President secured visibility for the report and was determined to advance a Limits of Growth agenda, his efforts brought little result. Within the Commission, Mansholt's ideas were not taken seriously. "Sicco, are you becoming a hippy?" asked Altiero Spinelli, commissioner for Industrial Affairs and Research. While some of these reservations might have stemmed from Mansholt's alarmist style and his increasingly radical socialistic solutions, throughout the 1970s, economic growth would continue to dominate the Community's political imaginary and agenda.[30]

This continuity also characterized discussions on research policy. The debate of the "limits" and the new sensitivity toward the adverse effects of human activity had increased public interest in science and technology. After all, it was scientific progress that had made the large-scale exploitation of natural resources possible, and that seemed to lie behind not only the historically unforeseen rates of economic growth but also its nasty side effects. However, this issue mainly remained a concern of academic circles and was related to the objectives and methods of research and research policy, not so much science itself. Furthermore, the general contention was that the harmful effects of scientific activity could be fixed by introducing more efficient and cleaner technologies, thus by introducing more science. Technology assessment could also be used to identify negative externalities. There was thus very little change to the policy frame reflecting modernist confidence in the inevitability of progress and the science-led process to economic growth, yielding broader social benefits.[31] Among the general public, science also continued to enjoy prestige and appreciation. In 1972 the DGRST, concerned about the public protests against directions taken by investment in scientific research, which it had observed in other countries, commissioned a survey on public attitudes toward science in France.[32] The result of the survey showed no signs of a confidence crisis.[33]

In this context it seems logical that the 1970s revisions of science policy were partial and mainly motivated by economic concerns. The first revisions were made in the Unites States, where the balance-of-payment problems and later inflation and recession turned attention to the economic payoff of R&D. At the same time, social unrest produced a demand for R&D activity in new sectors, while triggering, together with the Vietnam War, sharp criticism of the military-industrial complex. These pressures were soon reflected in the government's action: the agencies that had dominated the US postwar science policy, dictated by ambitious "strategic objectives"—the Department of Defense, the Atomic Energy Commission, and the National Aeronautics and Space Administration—were rapidly subjected to radical budget cuts.[34] Ivan Bennett, the deputy science advisor to the president in the Johnson administration, summarized the new spirit of the US science policy in 1968:

> Science…can no longer hope to exist, among all human enterprises, through some mystique, without constraint or scrutiny in terms of national goals, and isolated from the competition for allocation of resources which are finite.[35]

By the early 1970s, the US science policy had entered a new period of a decisively colder political and economic climate.

Important policy revisions were also conducted in Western Europe. In Britain, a report from the UK Science Policy Council[36] of 1967 gave clear indications of new rigors of making science policy, which prompted John Walsh to conclude in *Science*: "As in the United States, an era has apparently ended during which the science budget was boosted annually almost as an act of faith. What Vietnam has meant to science in the United States, a limping economy means to British science."[37] Fundamentally, the main reason for the British government to consider policy revision was the same as in America: disappointment with the results of substantial R&D expenditure, and a desire to see greater returns of investment.[38] "In Britain, profitability is now openly proclaimed to be a major consideration in government support of research," *Science* magazine wrote in September 1969.[39] There seemed to be a rather common feeling that promised payoffs from science had never really materialized and that British

technological prestige, even in the fields where the country had traditionally been in the lead, such as aerospace, had now waned. What followed
was a vivid discussion on the effectiveness of government-sponsored civil
research and a decidedly more careful and selective use of public funds.[40]
In the early 1970s, British science policy moved into a more utilitarian
direction, while funds for basic research in universities leveled off after a
decade of steady increases.[41]

Similar trends were visible in France. In his study on the history of the
CNRS, Denis Guthleben recognizes a rupture in French research, following the events of 1968 and the subsequent departure of President Charles
de Gaulle. In addition to the end of the generous budgets granted by the
Gaullist government anxious to make science one of the main pillars of the
national grandeur, this break with the past was demonstrated by a slowing
down of the strong national programs that had generated a whole vein of
new research centers intended to gain international prestige. After 1969,
expenditures in areas such as defense, atomic energy, and space had been
leveled off or even cut back.[42] If in the 1960s, research in public institutions had been closely followed and strongly encouraged by the top
national decision-makers, in the subsequent decade, this engagement was
dramatically reduced.[43] More importantly, the change that Guthleben
describes as a "passage from the Golden Age to the Iron Age, without
transition" was sudden and radical. The "leveling off" in the science sector[44] had brutal and long-lasting effects for France, even more than for
Britain and the United States, because in France, the economy had been
rather strong and the reliance of the scientific community on government
policy and funds had been more direct. Moreover, the change was not balanced by an increasing investment in research in French industry. On the
contrary, the industry had started to tighten its research budgets already
before the government announced these cuts.[45]

Both in France and Britain, the turn of the decade brought administrative changes that pushed science policy and economic policy closer
together. In France, the Ministry of Science and the Ministry of Industry
were merged in 1969 creating a Ministry of Industry and Scientific
Development. In the following year, the British conservative government
merged the Ministry of Technology with the Department of Economic
Affairs and the Board of Trade, creating the Department of Trade and
Industry. This came a few months after the UK Science Research Council's
announcement of a reorganization designed to emphasize research and
training related to industrial needs.[46] According to Wayne Sandholtz,

"[t]hat move symbolized better than anything else the marriage of science and technology with economic objectives."[47] It seems that throughout Western Europe, profit now played an increasingly central role in science policy. D.S. Greenberg, writing from London to *Science* magazine, concluded in October 1969 that "science policy-making in Western Europe has entered a utilitarian period that leaves little room for those who would argue in behalf of science for the sake of science. ...[T]he Common Market countries hold endless meetings on all the good things they might someday do together technologically but, throughout, the emphasis is on utilitarian projects." This message was also confirmed by Anthony Wedgwood Benn, the British Minister of Technology: "We have come to the end of those days when any project, sufficiently big and spectacular, could almost automatically expect to win approval and the funds it needed... The era of technomania is passing."[48]

The attempts of European governments to orient research to attain more precisely defined economic and social goals coincided with a reassessment of the "linear model" of innovation. Already around the mid-1960s, American economists started to question this rather simplistic differentiation between basic and applied sciences, on which much of the postwar thinking on research policy had been based. Once again, a major channel for the transmission of new perspectives on science and economy was the OECD. In 1971 a group of eminent science policy experts produced an influential report that captured the spirit of the times. This so-called Brooks report, named after the group's chair Harvey Brooks, the dean of Division of Engineering and Applied Sciences at the Harvard University, recognized a major change in the overall climate in which science had been operating since 1945, and called for a radical reappraisal of the science policies practiced so far in the OECD member states. Describing the early 1960s as "'the naïve period' of understanding science in relation of economy," the Brooks report acknowledged that "[t]he relationship between economic growth and R&D investment turned out to be far more complex and less linear than originally supposed."[49] According to the Brooks Committee, "building up scientific capabilities was no doubt right, but basing it on the implicit assumption that research and development expenditures necessarily mark a directly proportional contribution to growth was unwarranted."[50] While holding to the idea that R&D contributes to growth, the Committee urged that this relationship should be perceived in the framework of a more complex process. There were many factors contributing to innovation, such as capital availability, fiscal policy, management competence, entrepreneurship, marketing skills, labor relations, general levels of education, and even culture and national psychology.[51]

These observations stemmed from the empirical evidence which pointed out that there were countries with high growth rates but little technological innovation of their own, and vice versa. A commonly cited example, clearly disputing the claim of Vannevar Bush that "a nation which depends upon others for its new basic scientific knowledge will be slow in its industrial progress and weak in its competitive position in the world trade,"[52] is Japan, which still in the 1960s imported most of its technology but also enjoyed rapid economic growth.[53] Indeed, the case of America was not much different. The United States reached its technological lead when it was still relatively weak in basic science.[54] On the other hand, Britain had invested heavily in R&D but lagged behind in the international growth race.

The Brooks report also contended that the new political atmosphere marked a clear difference to the earlier years of public enthusiasm and political consensus, which had been directly translated into a dramatic increase of R&D spending. "The increase of resources was so rapid and so continuous," the OECD group recalled "that people were not far from regarding it as an inevitable phenomenon, virtually outside of the area of political choice."[55] The new decade began with an unexpected "disenchantment with science and technology" and made research policy a far more controversial affair that necessitated new priorities and painful political choices. However, what the Brooks report suggested was not a retreat from an economically driven science policy but rather an extension of the concept: "a new concept of science policy may well have to extend greatly the current boundaries of science policy as it is presently understood. It may be that the interactions of science policy with other policies will prove to be more important than its own mechanisms and objectives."[56] The Brooks report, heavily influenced by the discussions of the preeminent Club of Rome,[57] soon became the general reference point in the European and American debates on science policy.

The changes in the discourse and policy of science and technology served to calm European panic about the technology gap. Criticism of the generous public R&D spending that in the 1960s had mostly favored large demonstration projects and the doubts about the accuracy of the linear model of innovation reduced the political leverage of anxious comparisons between Europe and the United States. At the same time, the

economic recession that hit America and the 1967–1977 stagnation in the
volume of the US government-sponsored R&D made it seem as if the gap
was narrowing. Even though in absolute terms the US expenditure on
research continued to be the highest in the world, the differences between
the American record and those of Western Europe and Japan were rapidly
diminishing. For instance, in high-technology industries (defined by the
US National Science Foundation as industries having 25 or more scientists
and engineers per 1000 workers and spending at least 5 percent of sales on
R&D), US performance declined in relation to that of its competitors.
While in 1959, about 79 percent of the world markets were held by the
largest American firms, in 1978, their share was only 47 percent. In addition, between 1960 and 1980, American productivity growth, as measured by the output per employee in manufacturing, was slower than in
many other countries.[58] Toward the turn of the decade, these and other
similar figures were increasingly taken to show to the Europeans that the
gap was far less threatening than previously judged.

There were also some other, and more explicitly political, reasons for
the diminishing attraction of the gap. After the departure of President
Charles de Gaulle in France, from where much of the gap debate originated, the French drive for responding to the American challenge lost
some of its imperative. De Gaulle's successor, Georges Pompidou, was less
worried about France's independence and assumed a more conciliatory
approach to the United States,[59] which was also demonstrated by a new
spirit of scientific cooperation between the two countries. In 1972 John
Walsh wrote in *Science* that the French were already yawning about the
gap.[60] While the gap did not completely disappear from the European
debates on science, in contrast to the previous years, it was now embraced
with much more caution. In fact, it was now America's turn to worry
about declining competitiveness in the world markets. Ebbing out in
Europe, the gap debate gained strenght in America where the idea of a
reverse technology gap started to take hold.[61]

Furthermore, if the gap controversy had ever really been fueled by Cold
War reasoning and Soviet efforts in the field of science and technology, in
the 1970s, references to the USSR scientific challenge practically disappeared from the discourse. From this point on, science policy in the East
and in the West seemed to follow radically different paths. While the
Western countries moved away from the big demonstration projects that
had been characteristic of the early Cold War years, and engaged in efforts
aimed at increasing the social accountability and economic impact of
research, in the Soviet sphere of influence, there were no signs of a serious

reconsideration of the prevailing practices. The cult of science, carefully built during the Stalinist rule and consolidated in the subsequent years on the foundation of success in sectors such as space, nuclear power, and high-energy physics, remained unchallenged. Public disapproval with expenditure on prestige technologies with little immediate social utility, if it occurred, was effectively muted. As before, military R&D dominated national programs, while the entire research system became increasingly inefficient and incapable of providing pioneering inventions or matching the development in Western Europe and the United States.[62]

In the 1970s, the concept of brain-drain, referring to the mass emigration of European scientists to the United States, was to follow a similar development as the debate surrounding the gap. By the turn of the decade, the phenomenon showed clear signs of abating. This can partly be explained by the absence of factors that originally created the flow of European researchers to American labor markets. In the early 1960s, the United States had been building impressive space programs, while Britain was reducing its defense establishment and Germany still had rather limited R&D expenditures. In this situation, Britain and Germany became the major sources of large scientific and technological emigration. Ten years later, however, several new factors, such as the cuts in the US research budget, the American social and campus unrest, and the Vietnam War, discouraged Europeans from going overseas.[63]

It is interesting that the fading of the gap and the brain-drain controversy coincided with mounting difficulties in international scientific cooperation. In 1971 the Brooks report noted that "[t]oday, the OECD countries are calling these joint undertakings into question. The programs of some organizations, and even the organizations themselves are being challenged…some countries have even withdrawn from the cooperative programs at issue. An evident uneasiness is spreading among government, scientific, and academic circles."[64] As an example the Brooks Committee mentioned the continued crisis of Euratom and the setbacks encountered in the European Space Vehicle Launcher Development Organisation (ELDO) and the European Space Research Organization (ESRO), which it saw stemming from divergent national science policies and the fact that governments tended to regard cooperation as an end itself without seriously discussing the various benefits it afforded to each of the participants. This was seen as worrisome, since in the coming years, due to growing specialization of instrumentation and skills, there would be increasing constraints on the independence of nations in many technological areas.[65]

Indeed, in the 1970s, the rise of multinational corporations started to shape the idea of international research. Starting in the 1950s, multinational manufacturing firms had begun to reorganize their production systems, and beginning in the 1970s, they strongly resumed foreign direct investment that had been reduced during the two world wars.[66] In 1971 Caryl P. Haskins observed that:

> the coming decade is sure to witness the continuing and dramatic evolution of another powerful force for international scientific and technical collaboration, one that could ultimately bring about an actual melting of scientific and technological resources among a number of developed nations: the further growth and proliferation of the great multinational corporations. …Their development – perhaps the most striking event in the whole history of scientific and technological collaboration among nations – introduces a factor wholly new in scale and in considerable measure new in kind.[67]

To an increasing degree, research activity was undertaken outside national institutions and out of the reach of the traditional policy instruments of the state. This said, in the 1970s and 1980s, the globalization of R&D in multinational companies still remained limited. As Geoffrey Jones has noted, firms operating in industries with high technological opportunities continued to concentrate the R&D activities in their home countries.[68]

The emergence of a new mode of research activity in the framework of multinational enterprises, the problems in intergovernmental research cooperation, the waning political enthusiasm for supporting basic science, and last but not least, the new uncertainty concerning the relationship between science and economy constituted important challenges to the plans for a common European research policy. An OECD report of 1972 summarized the situation:

> Neither the rate of growth in the resources allocated to research activities, nor the objectives on which these resources have been concentrated for more than twenty-five years past, can be regarded as a permanent feature. On the contrary, the period, which is now opening seems likely to be characterized by great uncertainties as to the ways and means of progress – or possibly of regression – of resources and the whole nature of new research programs.[69]

The plans for an EC research policy were now developed in a more challenging context, characterized by uncertainty, economic austerity, and a growing awareness of the negative externalities of science, technology, and economic growth. In the Western industrialized world, the late 1960s and the early 1970s can be described as a critical moment in which policy-makers were challenged to reconsider the core fundaments of research policy and to adapt to a transformed and increasingly complex ideational and political framework. It was a brief moment of opportunity to reconsider underlying political objectives and values and to reform national economic and research systems. Yet, more important than a break, was continuity. Global economic trends that had already been visible during the early 1960s, and that had elevated the utilitarian vision of research policy, grew stronger. Market liberalization, the sharpening economic competition worldwide, and rapid technological development continued to create pressure for governments. At the same time, the energy crisis, the monetary turmoil, stagflation, and the intensified North-South conflict effectively shifted the political attention away from the negative aspects of growth. The OECD member states, trying to tackle these new challenges and the growing pressures of the increasingly open global economy, soon lost interest in the debate on long-term problems associated with growth, welfare, and ecology.[70] Research retained its prominent role as a booster of economic growth and as a remedy to multiple social and economic challenges.

According to Lutz Raphael and Anselm Doering-Manteuffel, it is this parallel existence of the elements of constancy and evolution which underlines the status of the 1970s as a transition period in Europe.[71] During those turbulent years, the entangled existence of old and new was also the defining element in European science policy. The unfailing desire for growth, coupled with the new challenges of technological change and international economic contest, carried the utilitarian concept of research policy forward and cemented its centrality in European affairs.

NOTES

1. For instance, in the United States, the Ford and Carter administrations perceived damping inflation and cushioning recession as the first priorities of national economic policy. "The watchword of this new environment," writes Robert Collins, "was not growth but stability, especially price stability." Robert M. Collins, *More. The Politics of Economic Growth in Postwar America* (Oxford: Oxford University Press, 2000): 152–159.

2. Charles S. Maier, "Two Sorts of Crisis? The 'long' 1970s in the West and the East," in *Koordinaten deutscher Geschichte in der Epoche des Ost-West-Konflikts*, ed. Hans Günter Hockerts (München: Oldenburg, 2004),

49–51. According to Maier, the two other conflicts were the crisis of representation that lasted from about 1905 through World War I, while the second was the crisis of the world economic depression of the 1930s.

3. See also: Charles S. Maier, "'Malaise', The Crisis of Capitalism in the 1970s," in *The Shock of the Global. The 1970s in Perspective,* ed. Niall Ferguson et al. (Cambridge, Massachusetts and London, England: The Belknap Press of Harvard University Press, 2010), 25–48.

4. Reinhard Steurer, *Der Wachstumsdiskurs in Wissenschaft und Politik: Von der Wachstumseuphorie über "Grenzen des Wachstums" zur Nachhaltigkeit* (Berlin: Verlag für Wissenschaft und Forschung, 2002), 18.

5. Harold James, *Europe Reborn. A History, 1914–2000* (Harlow, UK: Pearson-Longman), 2003, 306.

6. Carole Fink, Philipp Gassert, and Detlef Junker, "Introduction," In *1968: The World Transformed,* ed. Detlef Junker, (with assistance of Daniel S. Mattern) (Washington D.C.: The German Historical Institute and Cambridge University Press, 1998), 2–3, 7.

7. Collins, *More,* 68–77, 86, 97.

8. OECD: *Science, Growth and Society. A New Perspective* (Paris: 1971), 32.

9. Ibid., 21.

10. Collins, *More,* 136–139. For a critical and insightful analysis of the rise of the environmental movement, see Frank Uekoetter, *The Greenest Nation? A New History of German Environmentalism* (Cambridge, MA: MIT Press, 2014), 101–111. Uekoetter challenges the simplistic narrative of a grand predestined awakening and explanations suggesting that the environmental movement emerged as a logical result of problems needing solutions.

11. Michael Bess, *The Light-Green Society: Ecology and technological modernity in France 1960–2000* (Chicago and London: The University of Chicago Press, 2003), 76–77.

12. Uekotter, *The Greenest Nation,* 78–79.

13. Benjamin Kline, *First Along the River. A Brief History of the U.S. Environmental Movement* (Lanham: Rowman & Littlefield, 2007), 84; Uekoetter, *The Greenest Nation,* 84.

14. Wolfram Kaiser and Jan-Henrik Meyer, "Introduction: International Organizations and Environmental Protection in the Global Twentieth Century," In *International Organizations and Environmental Protection. Conservation and Globalization in the Twentieth Century,* edited by Wolfram Kaiser and Jan-Henrik Meyer. (New York and Oxford: Berghahn Books, 2017).

15. The debate about the feasibility and the desirability of economic growth goes back to the late eighteenth century: the enquiry of classical economists into the physical feasibility of growth was followed by concerns whether growth could be sustained with the existing institutional frame-

work. To these considerations, the Club of Rome report added the physical constraint of environmental and ecological balance. Keith Pavitt, "Malthus and Other Economists. Some Doomdays Revisited," in *Thinking about the Future. A Critique of the Limits to Growth*, eds. H.S.D. Cole et al. (London: Chatto & Windus for Sussex University Press, 1973), 137–154.

16. Donatella H. Meadows, et al., *The Limits to Growth. A Report for the Club of Rome's Project on the Predicament of Mankind* (New York: Universe Books, 1972).

17. Collins, *More*, 140.

18. See also Matthias Schmeltzer, *The Hegemony of Growth. The OECD and the Making of the Economic Growth Paradigm* (Cambridge: Cambridge University Press, 2016): 246–266.

19. Ibid., 268, 273, 283–284.

20. Collins, *More*, 109–125.

21. Erzan J. Mishan, *The Economic Growth Debate. An Assessment* (London: George Allen & Unwin LTD, 1977), 25.

22. Steurer, *Der Wachstumsdiskurs*, 20.

23. Economic growth was recognized by the federal government as an explicit political goal only in 1967. In this context the coalition government also announced for the first time an objective of a growth rate of 4 percent, which was to be reached by the means of planning (*Globalsteuerung*). Ibid., 369–370.

24. Ibid., 370–374.

25. Ibid., 378, 382, 383, 389. The author's translation from German.

26. Uekoetter, *The Greenest Nation*, 98–99.

27. Johan van Merriënboer, "Sicco Mansholt and 'Limits of Growth,'" in *Europe in a Globalizing World. Global Challenges and European Responses in the "long" 1970s*, ed. Claudia Hiepel (Baden-Baden: Nomos, 2014): 319–342.

28. Steurer, *Der Wachstumsdiskurs*, 421–432.

29. Kline, *First Along the River*, 84, 99, 101.

30. van Merriënboer, "Sicco Mansholt," 334, 341.

31. Johan Schot and W. Edward Steinmueller, "Three Frames for Innovation Policy: R&D, Systems of Innovation and Transformative Change," *Research Policy* 47 (2018): 1556.

32. Daniel Boy, "Attitudes toward Science in France 1972–2005," in *The Culture of Science – How the Public Relates to Science Across the Globe*, eds. Martin M. Bauer, Rajesh Shukla, and Nick Allum (New York: Routledge, 2011), 39.

33. Ministère du développement industriel et scientifique. CCRST. Procès-verbal du 16 février (371ème réunion). 019920547, art. 6, Archives Nationales Fontainebleau (herafter ANF).

34. Jean-Jacques Salomon, "Science Policy Studies and the Development of Science Policy." In *Science, Technology and Society: A Cross-Disciplinary Perspective*, eds. Ina Spiegel-Rösing and Derek de Solla Price (London: Sage, 1977): 57; John Walsh, "OECD Report on Research System Says the Honeymoon Is Over." *Science*, May 30, 1975.

35. Quoted in William Bevan, "The Welfare of Science in an Era of Change." *Science*, June 2, 1972.

36. The Council of Science Policy, created in 1965, consisted of distinguished nongovernment scientists and science administrators. Its main task was to advise the Secretary of State for Education and Science.

37. John Walsh, "British Science Policy: The Case for Growth." *Science*, November 24, 1967.

38. John Walsh, "British Science Policy: After the Great Debate." *Science*, May 4, 1973.

39. D.S. Greenberg, "France: Profit Rather than Prestige Is New Policy for Research." *Science*, September 26, 1969.

40. John Walsh, "British Science Policy: A Crisis of Confidence." *Science*, November 5, 1971.

41. Walsh, "British Science Policy," May 4, 1973.

42. John Walsh, "French Science Policy: Problems of Leveling Off." *Science*, March 31, 1972.

43. Denis Guthleben, *Histoire du CNRS de 1939 à nos jours: une ambition nationale pour la science* (Paris: CNRS, 2009): 258.

44. In France, the proportion of science in the GDP fell from 2.4% in 1968 to 2.1% in 1971 and 1.7% in 1976. Jean-François Picard, *La République des savants. La recherche française et le C.N.R.S.* (Paris: Flammarion, 1990): 252.

45. Walsh, "French Science Policy," March 31, 1972.

46. D.S. Greenberg, "Britain: New Emphasis on Industrial research." *Science*, October 24, 1969.

47. Wayne Sandholtz, *High-Tech Europe: the Politics of International Cooperation* (Berkley and Los Angeles: University of California Press, 1992): 63.

48. Quoted in Greenberg, *Science*, October 24, 1969.

49. OECD, *Science, Growth and Society*, 41.

50. Ibid., 41.

51. Ibid., 42.

52. Quoted in Donald E. Stokes, *Pasteur's Quadrant. Basic Science and Technological Innovation* (Washington D.C.: Brookings Institution Press, 1997): 23.

53. OECD, *Science, Growth and Society*, 42.

54. Stokes, *Pasteur's Quadrant*, 23, 94.

55. OECD, *Science, Growth and Society*, 38.
56. Ibid., 12.
57. The ten authors of the Brooks report overlapped to a large degree with the network around King and the Club of Rome. Schmelzer, *The Hegemony of Growth*, 263–265.
58. Bruce L. R. Smith, *American Science Policy since World War II* (Washington D.C.: The Brookings Institution, 1990): 102–104.
59. Keith Pavitt, "Technology in Europe's Future." *Research Policy*, 1 (1971/1972), 210–273.
60. Walsh, "French Science Policy," March 31, 1972.
61. Harvey Brooks, "National Science and Technology Policy and Technological Innovation," in *The Positive Sum Strategy: Harnessing Technology for Economic Growth*, eds. Rolf Landau and Nathan Rosenberg (Washington DC: The National Academy of Sciences, 1986), 132.
62. Paul R. Josephson, "Soviet Scientists and the State: Politics, Ideology and Fundamental Research from Stalin to Gorbachev," in Margaret Jacob, *The Politics of Western Science*, 1940–1990 (New Jersey: Humanity Press 1994), 119–124.
63. Pavitt, "Technology in Europe's Future," 220.
64. OECD, *Science, Growth and Society*, 48.
65. Ibid., 50–51.
66. Geoffrey Jones, "Multinationals from the 1930s to the 1980s," in *Leviathans: Multinational Corporations and the New Global History*, eds. Albert D. Chandler Jr. and Bruce Mazlish (Cambridge: Cambridge University Press, 2005), 102–103.
67. Caryl P. Haskins, "Science and Policy for a New Decade." *Foreign Affairs*, January 1971, 240.
68. Jones, "Multinationals," 102.
69. Quoted in John Walsh, "OECD: Report Sees Closer Links between Research, Social Objectives." *Science,* April 14, 1972.
70. Schmelzer, *The Hegemony of Growth*, 285–287.
71. Anselm Doering-Manteuffel and Lutz Raphael, *Nach dem Boom. Perspektiven auf die Zeitgeschichte seit 1970* (Göttingen: Vandehoeck & Rupprecht, 2008): 42.

CHAPTER 6

COST: Distraction or Progress?

Initially, the reappraisal of research policy, both in Western Europe and in the United States in the late 1960s and the early 1970s, seemed to have little direct impact on the debates in the European Community (EC). In 1968 and 1969, the arguments for an extension of the Community's competences largely followed conventional lines: research was seen as important because it held out the promise of growth. For the Community decision-makers, it was the imperatives of economy, the allure of "more," that was regarded the best justification for new initiatives in the field of research. However, in spite of this continuity, and the affirmations the EC research ministers had made in the Council meeting of October 31, 1967, the following years brought little progress in this domain. In fact, events took an unexpected course that did not quite correspond with the initial schemes: instead of a substantial increase in the powers of the Community institutions, the focus shifted to intergovernmental cooperation à la carte.

This orientation was cemented by the establishment of COST (European Cooperation in Science and Technology), a loose and broad array of research projects bringing together nineteen Western European countries.[1] With its wide geographical scope, weak representation of the Commission, parity between member and non-member states, and the provision of the secretariat by Council and not the Commission of the EC, COST marked a fundamental break with the earlier proposals for activity, with a distinctively more Community-centric design.[2] COST was an unwanted child, a lukewarm compromise that actually pleased few of its

© The Author(s) 2020   147
V. Mitzner, *European Union Research Policy*, Europe in Transition:
The NYU European Studies Series,
https://doi.org/10.1007/978-3-030-41395-8_6

architects. For the Commission, it was an unfortunate mishap in its attempts to create new activities in the framework of the Six, while for most member governments, COST represented a mediocre outcome of political bargaining with little significant substance. Only the French government seemed to be relatively happy with the result. The French leaders saw the initiative as another victory in the Gaullist crusade against supranationalism and the growing ambitions of Brussels.

The decisive event in the history of COST and the evolution of the Community's debate on research policy more generally was the second British application for membership in May 1967, and perhaps, even more importantly, the French veto of the following December. From this point onward, the disagreements over the inclusion of new member states in the EC came to dominate the discussion on research. The urgent need to overcome the political deadlock resulting from the veto and the fact that technology was identified as an area where progress in Community enlargement could be sought, transformed the outlines of research policy into schemes of intergovernmental cooperation between the Six and the applicant countries. From now on the principal question was not whether the EC should have a research policy and what form this policy may take, but rather what kind of framework would allow the involvement of Britain and also perhaps some other non-member states. After 1967, research was subsumed by the full-blown struggle over the British membership, and in this way, it was rapidly subjected to short-term solutions, mostly shaped by other political objectives.

This chapter shows how the negotiation over the EC enlargement and the distribution of political power between national governments and the Community institutions conditioned the development of the Community's activities between 1967 and 1971. This debate involved two major disagreements: the first was temporal, concerning the stage and modalities of the participation of non-member states in the workings of the Community, and more precisely the question of whether or not the Six should first agree on the basic principles of their new activities among themselves before engaging in discussions with other European countries. Here a common view was that the earlier the other governments were allowed to influence the works, the weaker the Community element in the resulting activity would be. The second disagreement was spatial and revolved around the question of the geographical scope of the prospective action. Also, this was seen to determine the Community's future role: the more countries there were involved in the joint projects, the more flexibility would be required and the less political weight the Six and the Brussels institutions would have. After the request for participation from Yugoslavia, Greece, Finland, and Turkey, COST threatened to become diluted into a vague umbrella organization covering a large geographic area. Through this prospect, COST represented a challenge for the EC to determine its relationship to a wider Europe where it sought a profile as a natural core.

As a reaction to the sudden veto of President de Gaulle to the British membership application, in January 1968, the Dutch and Italian governments refused to participate in the subsequent meetings of PREST—a working group, also called the Maréchal committee, that the Council had mandated to examine the opportunities for a common research policy. This led not only to cancellation of the reunions scheduled for the scrutiny of the already completed reports from the sectorial working groups,[3] but also to nearly a year-long standstill in the Community's work in the domain. The French maneuver especially angered the Dutch government who, long disillusioned by the Gaullist European policy, decided to refrain from the examination of any new areas of the Community's activity without the participation of Britain from the outset.[4] Essentially, the boycott was meant to put pressure on France regarding the enlargement: the Dutch, wary of remaining locked up in a limited and protectionist continental bloc dominated by France and Germany,[5] had been the staunchest supporters of the British EC membership. The Hague saw a real danger in the United Kingdom not accepting the conditions for cooperation laid down by the Six. From the Dutch perspective, developing new activities within the EC at this point when the British accession was once again relegated to an unknown future, could only lead to an increase in the distance between the Community and the applicant countries. Therefore, if new activities were launched, the British and eventually also the other countries wishing to join should be invited to participate from the outset. Similar views were held in Rome, where Foreign Minister Fanfani clarified his government's stand to State Secretary Brown in late December 1967:

> [T]he position of the Italian government could be defined with two phases: (a) it would not have taken any initiative that could have made the British participation [to the Community] more difficult or distant, (b) it would not have carried out any action that could have destroyed anything that already exists.[6]

Science and technology was becoming a sensitive issue in the struggle over the enlargement.

Obviously, the Dutch and Italian governments were encouraged by the outspoken willingness of Britain to develop technological cooperation within a framework of the enlarged EC. In his Guildhall speech of November 1967, Prime Minister Harold Wilson had stressed that the discussions about technology in the European framework should begin immediately and proceed simultaneously with the talks about the British accession to the Community.[7] Even after the French veto, London stuck to its "technology card"[8]: in early 1968, the United Kingdom was wary of giving the impression that "technological collaboration with Britain was something the Europeans could achieve without putting themselves to the inconvenience of letting us into the Community." This was not easy. After France's second refusal, Britain's room for maneuver was limited. It was walking "a rather narrow path between, on the other hand, doing nothing, and on the other, falling into this trap. The standard French line of talk was that it was perfectly possible for France and Europe to collaborate with the United Kingdom technologically without any institutional ties."[9] However, in a situation where the deteriorating domestic economy in Britain weakened the prospects for the entry into the Community, underlining scientific and technological assets only gained more importance. A Foreign Office note summarized in March 1968: "In these circumstances, technology is one of the main fields in which we might still hope to make substantive progress towards closer collaboration with the Community."[10] In 1968, Britain's line became more flexible, while London showed an increasing readiness to accept alternative arrangements that were not necessarily restricted to the Community framework. A good example of this orientation is the gas centrifuge project between Britain, Germany, and the Netherlands, launched in November 1969.[11]

But for London, besides easing the entry to the EC, discussing research and technology with the Six also had another strategic advantage: continuously assessed within the national governments and the Community institutions, the various propositions of scientific and technological cooperation took the focus away from other European initiatives that the British considered undesirable. The debate on research thus prevented the Six from moving too fast in areas that the British actually considered more vital and that they truly wanted to influence.[12]

Britain's continuing interest in research and technology and the frustration over the French behavior on the enlargement issue were not the only reasons for the Dutch to block PREST's work. In February 1968, the Dutch Foreign Minister Joseph Luns informed the British about his dissatisfaction with the development that was taking place in the Community. He especially found astonishing "how quickly the Community and the Commission got their grips on any subject which seemed to show promise of progress, and turned it into a Community topic even though it lay quite outside the Treaty of Rome." A good example of this was the Luxemburg resolution of October 31, 1967, on science and technology. In Luns' view, that had been a resolution made by the member states in their national capacity, but the Commission had immediately seized it. Disappointed, he had told his colleagues not to attend the following expert meetings and also tried to prevent the Italians, Belgians, and Luxembourgers from going. However, only Italians responded to his plea,[13] and with a much less resolute attitude. Sarcastically, Luns commented that Belgians and Luxembourgers "couldn't keep their fingers out of anything labelled 'Community.'"[14] Apparently, the Dutch government was not as willing to promote integration for its own sake as some of its partners, and disliked the Commission's eagerness to extend its own competences. Moreover, the Netherlands did not necessarily perceive the EC as the ideal framework for new European activities. In October 1967, Alan Smith from the Foreign Office noted:

> I imagined that the Dutch intransigence might be wholly political in origin, but when I spoke to the Dutch delegates to the OECD meeting...they put a rather different slant on things. They said they were thoroughly fed up with repeating in Brussels discussions, which had already taken place in the OECD. On science and technology it was obviously of more interest to discuss things with the British and Americans present–as they were in OECD–than in the restricted context of the Six. They were busy men, they said, and had no liking for wasting time. These remarks came at the end of the singularly dull OECD meeting, and were made with heartfelt sincerity.[15]

There was thus sincere frustration in the Hague with the tedious debates within the Community. The OECD was perceived as more efficient and interesting arena for pursuing cooperation with the most technologically dynamic Western countries. This explains why the Dutch were ready to sacrifice the Community's work on research in the pursuit of their political objectives of enlargement.

Keen to progress with the work launched by the Luxembourg Council, the other EC member states showed little understanding for the Dutch and Italian boycott. The attempts of Luxemburg and the Belgium to make the Dutch reconsider their position[16] were followed by pleas that the French and the German governments presented in COREPER in February.[17] But equally disappointed with the interruption of PREST's work was the Commission: on February 7, 1968, it addressed the Council president in a letter calling for an immediate reprise of the working group's activity.[18] A week later the Commission informed the Permanent Representatives about its intention to continue studies related to questions that the Maréchal group examined as well as about its decision to use its right to make propositions in case the working group would not soon resume its activities.[19]

It is interesting that at the very moment when the work of PREST came to a halt due to the disagreement over the enlargement, research was presented as a solution to the political deadlock on Britain's entry. Early in 1968, a number of proposals were tabled for organizing relations between the EC and the countries that had applied for membership, most notably Britain. An important aspect of these proposals was cooperation in the fields that were not covered by the Treaties.[20] Paradoxically, at this point the fact that the Community was lacking a proper treaty basis in research became an asset for the advocates of integration. Research could serve as a bridge between the EC and the applicants; it would be used as a "neutral" territory where rapprochement was easier than in sectors subjected to the Community legislation. Moreover, due to the continuing British interest in developing technological cooperation with the Six, and the broad consensus in the EC on the advantages of involving Britain in the Community's activity, research and technology was widely perceived as a good candidate for a field where interim solutions could be sought. A memorandum of the Auswärtiges Amt reveals this desire: "In our efforts to increase the integration of UK with the European Communities and their Member States, the field of science and technology plays an important role."[21] Consequently, various suggestions for cooperation were submitted between January and March by the Benelux countries[22] and Germany.[23]

The wide array of proposals notwithstanding, making progress was not easy. The Council meeting of March 9, where the different documents were discussed, ended with a disagreement. At this point the main bone of contention was the way in which the non-member states would take part in the Community's work. Whereas the Dutch called for the organization of a ministerial conference as soon as possible to map out the contours of the new activities the Six and the third countries could have together, the

French refused the idea of examining these questions in a framework larger than the EC. The French Foreign Minister Couve de Murville remained particularly uncompromising: for France, he stressed, the resolution of the Luxemburg Council of October 31, 1967, represented the optimal solution, and hence the only acceptable way to proceed was to go on with implementing these decisions. Only when the Maréchal group had finished its work, could the Six embark on a discussion in a wider geographical context.[24] This attitude antagonized the Dutch who after the Council started to speculate about leaving France outside the proposed conference.[25] It came down to the Belgian Foreign Minister Harmel to talk to his Dutch counterpart into smoothing things over.[26]

At the root of the French inclination to stick to a minimalist interpretation of the Luxemburg resolution and their stubborn opposition to the idea of letting non-member states, especially Britain, to influence the planning of new Community activities, was a fear of losing control over the development of the EC. For France the British involvement presented a danger that the Community could be steered in the opposite direction that the French desired. While basically accepting the idea of extending cooperation to other European countries, the French wanted to make sure that this cooperation took place under conditions set by the Six.[27] But this approach also stemmed from a more general change of attitude toward a common research policy. Since 1965, the French interest in multilateral activities in the field had decreased. Reflecting the tightening budgets and political turbulence at home, the priority was now given to cost-effective and concrete bilateral projects deemed to serve the national interest.[28]

The French refusal to discuss scientific or technological cooperation with non-member states before the Maréchal group had submitted its report, and the rebuttal of the Dutch government to participate in the meetings that would have allowed the work of the group to be finished, created a political impasse. However attractive research cooperation appeared as a way of strengthening the relationship with London, and however frustrated the other member states[29] and the Commission[30] grew, in the spring 1968, the prospects for a common research policy, be it with Britain or without, appeared rather dim. Furthermore, the British case had also suffered as the result of the Labour government's announcement to withdraw from various existing projects of European technological cooperation.[31]

It was only by early November when both the French and the Dutch adopted a more conciliatory approach: in the Council of November 4 and

5, the French delegation presented a series of concrete suggestions for a commercial arrangement and technological cooperation with Britain. In addition, the Dutch Foreign Minister Luns, while underlining that the Netherlands would still not accept the French view, said that the attitude of his government was not immobile. The Hague could approve the Maréchal group to carry on its work only if it was assured that a multilateral conference, bringing together all European countries interested in technological cooperation, would later take place.[32] One month later the COREPER, after many hours of intensive discussion, reached an agreement on a compromise proposal to be presented to the Council.[33]

On December 10, the Council decided to resume the Community's activity in research policy along the lines that COREPER had suggested. The deadline for putting the decisions of the 1967 Luxembourg Council into effect was postponed to July 1, 1969, and the Maréchal group was asked to submit its report by March 1, 1969. Furthermore, the Council would address propositions of cooperation with other European countries, especially those that had applied for membership.[34]

The new willingness of the governments to make concessions stemmed from two sources. One of them is identified in the general report of the Commission: "The resolution of 10 December…settled a very serious difference of opinion between the Member States, but one which was basically only a matter of procedure, namely, concerning the best approach to follow in implementing a joint policy in new fields of scientific and technical research, or even in any new field. It may be asked, after the event, whether the increasing difficulties experienced in almost all fields of scientific and technical co-operation at a European level did not themselves help to close the ranks."[35] While the crisis of Euratom pushed for creative solutions in a more comprehensive research policy framework, the simultaneous problems in other European organizations underlined the evils in strictly intergovernmental cooperation. For the Commission, they served as a clear demonstration of the need for a more comprehensive and integrated institutional context. Altiero Spinelli, the enthusiastic Commissioner with responsibility for research, frequently presented this view. Only a common research policy within the EC would provide a coherent vision of the ensemble of activities that he deemed necessary for success.[36]

A brilliant case for failures in European intergovernmental cooperation was ELDO, The European Launcher Development Organisation, created in 1962 by six European countries[37] and Australia to develop a European space launcher. ELDO was a British initiative that stemmed from London's need to share the increasing costs of Blue Streak, Britain's own ballistic missile program, and to show greater European commitment in the context of the United Kingdom's first application to join the EC. But developing a joint European rocket program proved difficult. After a sluggish start, a number of teething problems appeared to disturb the cooperation, which very soon led to skyrocketing costs.[38] The subsequent attempts to reduce expenses then of course provoked political disagreements and open discontent, especially on the part of Italy and Belgium, the two countries on which most of the burden of budget cuts fell. By 1968, the days of the organization finally seemed to be numbered when the British government announced that it would not accept any new financial commitments. However, it was only after the failed test flight of Europa II in November 1971, that the member states concluded that a fundamental reform of the organization was unavoidable.[39] The situation was not much better in ELDO's sister organization ESRO, the European Space Research Organization, also set up in 1962.[40] In 1968 it became clear that the organization and the industry responsible for building ESRO's satellites had clearly underestimated the costs of their projects. This was particularly inopportune at a moment when many Western European states were revising their research policies and shifting the focus from science to applications while scaling back their financial commitments at the European level. These difficulties were aggravated to the point that in 1970, the ESRO was close to dissolution.[41]

Speaking at a conference in Grenoble in April 1970, Research Commissioner Fritz Hellwig said that after a wave of optimism that during the twenty previous years had translated into various attempts at scientific cooperation in Europe, there was now a period of reflection. This meant not only a reconsideration of the achieved results but also a search for a better system, "perhaps even a permanent principle of a genuine European research policy."[42] While supporting the efforts of the Council of Europe and the OECD to coordinate the dozens of international organizations dealing with research, he was convinced of the need for a more general policy for the EC that would go beyond supporting individual grand projects.[43] Commissioner Spinelli's Cabinet Chief Christopher Layton was also able to convert even the most discouraging European experiences to plausible political arguments to support integration. He too regarded the failures of intergovernmental projects as a reason to focus on the EC. In 1969 Layton wrote that "the present chaos of European organizations for science and technology cannot be allowed last much longer. Ad hoc bilateral projects, duplicated organizations, loose cooperative arrangements which fall apart at the seams with depressing regularity, absence of parliamentary control, may be excused as the first faltering steps of a child learning to walk. But it is time now for Europe to start growing up." His suggestion was a unified supranational community of which Britain would also be part.[44]

Whether the Commission's case of the evils of intergovernmental cooperation was convincing or not, the EC member governments were struggling with political pressure stemming from the more general impasse of the Community project. As a consequence of the prolonged row over EC enlargement, the entire Community was reaching a point where its further development was no longer possible. This worry also struck the French: according to N. Piers Ludlow, there was a growing awareness within the Gaullist government that those same interests that Paris saw being threatened by the Community's expansion could also be endangered by political standstill of the Community caused by non-enlargement.[45] The customs union had fully entered into operation in the previous July, and making it work as intended necessitated decisions that the French did not want to jeopardize by the disagreement over the British question.[46] It is also possible that the deterioration of the economic and financial situation in France in 1968 prompted the French to make concessions.[47]

The political pressure of the deadlock situation also had an impact on the Dutch who found themselves increasingly isolated. Even the support they received from London was weakening. Already in May 1968, a British observer noted: "The Dutch have been holding up the work of the Maréchal committee. It is in our interest that they should continue to do so; but if they are beginning to find their present isolation intolerable, it would be better not to force the issue."[48] And London didn't force it. Britain welcomed the accord of December 10, although first with caution. In particular the British press appeared satisfied with the outcome of the Council, which was described as a step in the right direction and a sign of a new flexibility on behalf of the French government.[49]

Most worried about the political impasse, however, was Germany. During 1968 the Bonn government made renewed attempts to find a solution to the enlargement dilemma. The liberal-democratic government was not only emphasizing the *Westpolitik* to counterbalance their overture toward the East—the Germans had also strong economic interests in overcoming the commercial divide between the six EC countries and the seven members of the European Free Trade Association (EFTA). Although in the 1960s German exports to the EC members had grown faster than they had to other European markets, at the end of the decade, EFTA still accounted for 23 percent of all Germany's exports. Through the removal of obstacles between the two trading blocks, the Germans hoped to regain some of their former markets in the EFTA area while preserving a strong position within the EC.[50] Bonn also regarded technological cooperation as an important means of strengthening the bonds to the Seven. A telegram sent from Bonn to the German representation in Brussels, prior to the December 1968 Council meeting, asked the German delegates to strive for an agreement practically at any price: "The most important goal of the current efforts is to unblock the work of the Maréchal group... Any formula on which the French and the Dutch can agree is acceptable to us."[51] Technology was included in the two sets of proposals on arranging the relationships between the Community and Britain that the Federal Republic presented in May and September 1968.[52]

But Germany's drive for revitalizing the EC's plans for research policy was not motivated by the trade and the desire to engage with Britain alone. While elsewhere in Europe, the fears of the technology gap had been somewhat muted, in Germany the economic recession of 1966/67 created the impression that the Federal Republic was falling behind.[53] An internal document of the German administration underlined the pressure of the gap and explicitly tied it with the initiatives for EC research policy in autumn 1968: "However, technological progress continues. It ignores the disagreements among Europeans. With every month we lose, it becomes more difficult to catch up with Europe's backlog in technology sector, on which it depends, and where we want to make more progress through more intensive European cooperation."[54] These considerations reflected views propagated in the German media. In 1969, *Der Spiegel* published a large article about the problem that it saw in the worsening international position of German technology and science. Although in terms of foreign trade Germany was doing well, up to 80 percent of its exports consisted of *Technik von gestern* [technology from yesterday] such as cars, machines, and steel and chemical products. The future sectors were mainly in the hands of American industry that dominated not only electronics (IBM produced more computers in one week than the second biggest German industrial enterprise Siemens AG could make in a year) but also sectors where Germany had once been the leader, like pharmaceutics and aerospace. Now it seemed that the most important innovations were actually foreign imports. The "brain drain" was another worry: according to *Der*

Spiegel, of the large legion of intellectuals that immigrated to the United States, German scientists formed a major part: "There is hardly a day that at least one academic does not turn his back to the Federal Republic."[55] Would this trend continue, the prospects for closing the gap were unlikely to improve.

At the same time, while in other large Western European countries the late 1960s and early 1970s witnessed a substantial leveling off of the government support for research, in Germany public funds for R&D continued to grow. In part, this difference can be explained by structural factors: in the United States, Britain, and France, the budget revisions mainly concerned large demonstration projects in military research or related fields that the countries had built up after the war. In Germany, by contrast, the outcome of World War II had limited the possibility of the government to embark on such massive and highly expensive undertakings, and consequently, in the time of austerity, there was no similar feeling of frustration with publicly funded research. This reduced pressure for major political reorientation. Helmut Trischler has argued that in Germany, the years of the Great Coalition and the Social-Liberal Government (1966–1982) represent a relatively closed period characterized by a remarkable homogeneity of the objectives, conceptions, and political implementation of science policy. The central role of the federal state in steering scientific research, the continuing enthusiasm about planning, modernization, and "big science" constituted the typical features of that distinctively uniform epoch.[56] Of course, as elsewhere in Western Europe, there were increasing concerns about the environmental and social impact of science and technology. But in face of these pressures, the government assumed a dual-track strategy: while continuing its generous financing of big science projects, Bonn extended its support to other domains of research deemed to provide responses to the new social and environmental challenges.[57]

The gap continued to also occupy a prominent place in the arguments of some other political actors in favor of greater European scientific cooperation. The Union des industries de la Communauté européenne (UNICE), an organization representing the industrial associations of the EC member countries,[58] allied with the Commission in support of a rapid conclusion of PREST's activity. In a letter addressed to the Commission President Jean Rey on July 16, UNICE expressed the concerns of the European industry about the delays in the application of the decisions taken in the Luxemburg Council of October 31, 1967.[59] This communication was followed by a memorandum on the problems and possibilities of a research policy within the Community.[60] UNICE also congratulated the EC for the Council decision of December 10, 1968, and reminded them of the continuing challenge of the gap: "In many sectors that are decisive for its future, Western Europe is lagging behind other economic powers, a delay that can only be bridged by large selective research and development projects."[61] Even if the recession and turbulence in the US and the simultaneous strenghening of European self-esteem had calmed down the gap debate, it was still widely used to promote European integration.

UNICE was not the only non-Community forum where the problems of the Maréchal group were debated: on October 18, a recommendation was adopted, by a vast majority, in the Western European Union (WEU)"[62] Council emphasizing the need to proceed with the PREST meetings and also invite experts from non-member countries. Although it is difficult to show the direct impact of all this activity on the actual decisions of the Council, the statement of a German observer concerning the WEU recommendation indicates that it was significant:

> The Assembly certainly helped to make it clear to the Member Governments that their agreement on the technology issue was inevitable within the European Communities, otherwise there was a danger that they would be under considerable pressure from their Parliament to consider other solutions outside of the European Communities.[63]

The possibility of developing cooperation within a framework other than the Community[64] seems to have pushed the governments toward a compromise.

> Finally, the European Parliament also repeated its worries about the gap and lamented about the delay in PREST's works. In a resolution of October 1, 1968 the Parliament considered with great concern, given the fundamental role of research and technology in economic development and the ever-increasing technological gap between Europe and the major technological powers, the delays in achieving the objectives laid down by the Council in these decisions of October 31 and December 8, 1967.[65]

The document mentioned the standstill of the activities of the group Maréchal, a "degradation" of the scientific cooperation both in the Community and in the European scientific organizations, and the "dangers" of a return to the national conceptions and programs. The Parliament called for an end to the dissipation of time, capital, and people that resulted from insufficient harmonization of national activities.[66]

External pressure, worries about a general deadlock of the Community, the activity of the German government, and last but not least, the continued fears of the technology gap thus pushed the Six to agree on the continuation of the Community's work on research. The first PREST meeting after almost a year of deadlock took place on January 7 and 8, 1969, in a good atmosphere. In April, the working group was able to submit to the Council a new report, the so-called Aigrain report, including forty-seven proposals for concrete action, and in June, the working group delivered a complementary document on the procedure for implementing these proposals.[67] This fresh start was highlighted by the appointment of a new chair, Pierre Aigrain, a French physicist, and Délégué général à la recherche scientifique et technique to follow André Maréchal in the lead of the group.[68] Aigrain was a scientific administrator of high caliber. Besides his long academic career, he had served as Scientific Director in Ministère des Armées (1961–1965) and Director des enseignements supérieurs in the Ministère de l'Éducation nationale (1965–1968). One might also note his American links: Aigrain earned his PhD in the United States where he also returned to serve as a visiting professor at the MIT.

The resumption of PREST's work did not mean that all disagreements were settled. At the governmental level, the question of the third country participation continued to divide opinions.[69] The Dutch and the Italians wanted to make a clear distinction between the four applicants for EC membership (Britain, Denmark, Ireland, and Norway) and the other countries, and grant the former a privileged status in the planned cooperation.[70] Yet, this differentiation displeased governments that had no immediate plans to join the Community but nevertheless wanted to be part of the EC's research activities.[71] Indeed, the popularity of the EC's plans for European scientific and technological cooperation was somewhat unexpected. All of a sudden, outsiders perceived the Community as a viable locus for European level activity. This reflected a broader trend: the completion of the customs union over a year ahead of schedule, the initial success of the common agricultural policy, and last but not least, the departure of Charles de Gaulle in France had generated a new sense of optimism about the future of the EC.[72] Despite the insecurities still prevailing about the enlargement, for many, the Community appeared as a dynamic enterprise, likely to assume new scope in the years to come. Joining research cooperation would be a smart way of engaging with this increasingly attractive partner.

In October 1969, the EC ministers decided to reach out to nine non-member states (Austria, Denmark, Britain, Ireland, Norway, Portugal, Spain, Sweden, and Switzerland) and offer other states the possibility to participle in certain actions at a later stage.[73] The reasons for giving the green light to Finland, Greece, Yugoslavia, Spain, and Portugal[74] were mainly political: after the explicit requests the Six received through diplomatic channels,[75] it was very difficult to exclude these countries without creating an image of a discriminant club of self-interested and indifferent states. Instead, COST could serve as a tool for détente. Apart from Finland, all these countries were ruled by authoritarian leaders, which prevented them from being considered as eligible candidates for EC membership. While excluded from the Community, they could nevertheless be drawn to the Western sphere of influence by creating ties through other forms of cooperation. This was important in the Cold War battle for hearts and minds, overshadowed by the threatening existence of the Soviet alternative.[76]

The Cold War considerations aside, the Community's new popularity had paradoxical consequences. With the multiplying expressions of interest across the continent, the Six were forced to reconsider the number of countries participating and actually the very nature of the activity envisaged. The decision to invite on equal basis all European countries that had expressed interest in participation, transformed COST into something very different from the original plans.

This development especially worried the Commission and some circles in the German and Belgian governments.[77] The German Foreign Ministry thought that the expansion of the circle of the participating countries would unnecessarily complicate the realization of individual research projects and obscure the objectives of a coherent and coordinated European science and technology policy.[78] For the Auswärtiges Amt, these objectives were twofold: research policy should, first, ease the accession of the candidate countries, and, second, promote integration in the framework of the enlarged Community. Opening the door to countries without realistic prospects for EC membership or at least closer relations with the Community would hinder the pursuit of these goals.[79] And last but not least, the invitation of countries, whose political room for maneuver was particularly dependent on the international situation, such as Yugoslavia, could carry an additional political loading and link the issue with the East-West conflict.[80]

The Commission was also concerned about its role in the initiative. The Brussels executive had offered to contribute to various COST projects through new research activities in Euratom's Joint Research Centre in Ispra. This perspective, however, did not please France. The disagreement lasted until the first COST conference on November 22 and 23, 1971, and the issue was still discussed in the Council meeting organized on the margins of the reunion. The French remained uncompromising: on the evening of November 22, the French Research Minister François-Xavier Ortoli announced that his government would not modify its position on the execution of the non-nuclear activities within the JRC and that the eventual participation of the EC should be decided on this basis and *cas par cas*. This firmness proved effective. A French memorandum summarized the result of the meeting:

> On the whole, our thesis has therefore prevailed since no new scientific competence has been recognized to the Communities apart from those which they hold under a strict interpretation of the Euratom Treaty and in application of the ECSC Treaty; the Commission, moreover, reluctantly accepted all the decisions of the Council, the manoeuvre it had tried turning against itself and its supporters.[81]

Once again, France had brilliantly succeeded in blocking a process that threatened to lead to an increase in the Commission's power.

The first COST conference approved seven projects in the following fields: information science, telecommunications, metallurgy, materials, and environment. The guiding principle of the proposals was flexibility; the member countries were allowed to select the projects in which they were most interested in participating.[82] Clearly, the French were happy with the result. In his address at the COST meeting, Ortoli stressed the value of a framework of nineteen participating countries, characterized by the *souplesse* of the procedures and the absence of the principle of *juste retour*. But the French could not ignore the fact that they were rather alone among their Community partners in their celebration of the outcome.[83]

In fact, the Five were ambivalent about the result and dismayed with France's attitude. The French stubbornness had put the EC in an embarrassing position, in which it had to explain its withdrawal from activities that already had been planned for the other countries. Only the British view, stressing pragmatism and flexibility, came close to that of Paris.[84] In the COST conference, the British Minister for Aerospace Frederick Corfield congratulated the Committee of Senior Officials for not having been too ambitious in embarking on new schemes. For him, the significance of the outcome resided exactly in the small-scale nature of the planned projects.[85] This of course reflected the changing trends in British research policy and the line Prime Minister Wilson was trying to follow also in European cooperation: withdrawal from big prestige projects and emphasis on pragmatism, profit, and the interests of industry were now the words of the day in London.[86] But it is also true that Britain, despite its enthusiasm to play the technology card, was never crazy about the Community-centric plans for research policy. This is an aspect of Britain's European policy that in the coming years became very clear, and that actually involves a paradox. Had the French not been trying to bar the United Kingdom from contributing to the Community's work on science in the fear of losing control over its contents? Yet, Paris found its best ally in London. Quite clearly, intergovernmental cooperation *à la carte* was exactly what also the British were looking for.

The loose design of COST was not the only reason why Britain appeared happy with the initiative. In 1971, with the membership negotiations opened, London had no reason to insist on its previous all or nothing approach while trying to use technology to facilitate its access to the Community. Already in 1969, a Foreign Office memorandum argued that there was little sense in "dragging our feet as a means of exerting pressure on the course of negotiations for entry into the EEC." Quite the contrary, "this tactic would be more likely to damage our prospects for entry." That was because "[o]ur record as a trustworthy European technological partner is not so pure that we can afford to be too clever by half in such matters. For us to seek to use our technological advantage to gain political concession would be resented and could easily be counter-productive." Moreover, even though in the sectors considered by the Aigrain group, Britain had some advantage over the Six, the situation was changing:

> [O]ur technological superiority...is a wasting asset. It is instructive that the Commission believes one of the fundamental changes between 1967 and today is that the technological gap between Britain and the Six has narrowed. The negotiation today would be much more a negotiation between equals in this respect. ...The fact is that we cannot for very much longer hope to be reckoned among the leaders in advanced technology across a wide front unless we can develop our skills on a European scale.[87]

In other words, in 1971 London could be satisfied also with an arrangement that did not necessarily guarantee membership in the Community.

The contrast between the COST and the earlier plans of a comprehensive research policy becomes clear when one takes a closer look at the Maréchal report of 1967. In the report the objective of a common research policy arising from detailed comparison (*confrontation*) of national research programs and budgets, and leading to harmonization and eventually specialization in national activities in research, was a recurrent feature. According to the document, the member states "should not be content to act in isolation through specialised projects making up a more or less diverse network of ad hoc agreements with different partners but...[should]... progressively put into effect a research policy valid for the entire Community."[88] By calling for "most flexible possible arrangements of cooperation," COST marked an explicit step away from the Community-centric approach to research policy.[89]

With nineteen members, also including technologically and economically weak countries such as Greece and Yugoslavia, COST could not serve as the main forum for joint research activity of the Community of Six, still worried about the consequences of the transatlantic technology gap. Indeed, as the work leading to the COST initiative advanced, the Community dimension was progressively weakened while the Commission found itself increasingly in the margins. In Brussels this did not go unnoticed: a report of the DG External Relations of July 23, 1971, warned that the already modest part played by the Commission in the preparatory work was in danger of further diminishing. Moreover, this development was being exacerbated by the behavior of the Brussels officials themselves: besides speaking in negative terms of COST, the Commission had appeared very formal and cold about the initiative. An attitude of this kind, the document warned, could bear serious consequences: "Most COST delegations are friendly towards the Commission but the coolness of the Commission makes them now look elsewhere – Council secretariat, national bodies – for certain tasks."[90]

By and large, the Commission's dwindling enthusiasm for COST was due to the character of the initiative that did not go along with the Brussels' vision of research policy. However, the Commission made little effort to steer the project into another direction. The compromising attitude[91] even went so far as to irritate some member states.[92] While the fusion of the Executives and the following delays with the reorganization of the general directorates might have contributed to this hands-off approach (in 1967 and 1968, the Brussels institution was not able to work as efficiently as it would have in normal circumstances),[93] pure political realism also played a role: as long as the enlargement question remained open and the Community's activity blocked, there was little sense in pressing on with proposals that everybody knew had poor prospects for success. In the discussions leading to the establishment of COST, the Commission's room for maneuver was limited.

Indeed, since the establishment of the Maréchal group in 1965, the Commission had had relatively little to say about the work on research and technology that had remained firmly in the hands of experts mandated by the national governments. It seems that the flexibility principle, for example, first proposed by the presidents of the PREST's special working groups,[94] was consolidated within the Committee of Senior Officials. This group, which largely composed of persons coming from various national ministries,[95] was established by the Council in 1970 to study the initiatives leading to COST.

If anything, the involvement of over 500 experts over the five-year period in the drafting of a new framework for European scientific and technological cooperation[96] shows the diversity and complexity of the channels of the transmission and formulation of policy ideas and concepts for integration. It also indicates the limits of the permanent Community institutions and the national governments in making choices between different policy options where special knowledge works as a central political resource. The fact that the proposals, presented by the working groups, especially the Committee of Senior Officials, were adopted without major modifications,[97] underlines the extraordinary ability of these experts to control and steer the discussion that was often of very technical nature.

At the same time, COST came into being as a political enterprise. Rather illustrative is the following statement of an ex-PREST member: "It was a bit of a political decision. In fact, several countries wanted to include Great Britain in these actions, but France did not want British participation to be understood as an advance admission to the Community. Great Britain was thus admitted on condition of admitting other countries. It was a compromise, which had nothing to do with research."[98] Pierre Papon, on the other hand, has argued that the French attachment to the *coup par coup* approach was a result of the excessive role of the Foreign Ministry and the Elysée.[99]

Despite the continuing worries about the technology gap and the concern for European economic performance, between 1967 and 1969, the design of the EC's activity in the field of research was strongly dictated by political considerations of other kind. In the struggle over the Community's enlargement, research policy fell victim to tensions between the Six and the subsequent efforts to find ways out of the political impasse. Linked to the enlargement issue, the work on a Community research policy soon changed its form, purpose, and direction. Opening cooperation with non-candidate countries only put the finishing touches to a project that now was developing along rather unintended paths. The irony of the situation is well depicted in a Foreign Office note from November 1970: "My own view is that we should go along with the idea that these additional three countries (Finland, Greece and Yugoslavia) should join in and any other countries who wish to do so. Now that we have 15, the more the merrier. …I agree with the view that we have got to the point now where the whole thing has become an absurdity and there is no reason why we should not make it more complete by the addition of extra useless countries."[100]

The invitation of nineteen countries to a cooperative framework initiated by the Community nevertheless demonstrated the remarkable territorial flexibility and inclusiveness of the process of European integration. As an extended Community initiative, it became an interesting hybrid that embraced several visions of European unity. Moreover, COST—whose European character was never contested[101]—arguably made the Community appear as an active initiator on the European scene and thereby increased the EC's visibility and credibility in the eyes of the others. Despite their many tensions and differences, in COST, the six EC member states acted as a unified group, which again increased the conception of the EC as a unitary actor. At the same time, with its ample membership and flexible modalities, the initiative added a new element of complexity, fluidity, and ambiguity to the process of European integration.

In the final analysis, the destiny of COST was, however, less determined by its unintended expansion into a loose collaborative fabric that barely served its initial purpose than by the rapid evolution of its political environment. In 1969, with the departure of Charles de Gaulle in France, the perspectives for the Community's future changed. Suddenly, the enlargement seemed possible again, and there was no more need to build temporary bridges between the EC and Britain, such as COST. But at that point it was too late to step back: the works on COST had already proceeded quite far, and the invitations to the third countries had been sent. Radical redesign did not come into question. The Community had no other choice than realize the project that fitted increasingly poorly with the transformed political reality. This would hardly satisfy the supporters of a more comprehensive concept of a common research policy.

NOTES

1. Austria, Belgium, Denmark, Finland, France, Germany, Great Britain, Greece, Ireland, Italy, Luxemburg, the Netherlands, Norway, Portugal, Spain, Sweden, Switzerland, Turkey, and Yugoslavia.
2. N.H. Aked and P.J. Gummett, "Science and Technology in the European Communities: The History of the Cost Projects," *Research Policy*, 5 (1976): 270–294.
3. Comité de politique économique à moyen terme: Groupe de travail Politique de la recherche scientifique et technique. Réunion du 30 novembre 1967, résumé des conclusions. CEAB2-3746/3, HAEU; Ergebnisbericht über die Sitzung der Arbeitsgruppe Politik auf dem Gebiet der wissenschaftlichen und technischen Forschung des Ausschusses für mittelfristige Wirtschaftspolitik der EWG am 30. November 1967 in Brüssel. B 35-82, PA AA; Ergebnisbericht über die Besprechung im BMwF am 14. Dezember 1967. B 35-82, PA AA. Réunion du groupe pour la recherche scientifique et technique. Extrait du p.v. spécial les 30, 31 janvier et 1er février 1968. BAC 118/1986-1395, HAEU; Parlement européen: Débats. Séance du mercredi 13 mars 1968, 56.
4. Comité des Représentants permanents. Projet de compte rendu sommaire de la réunion restreinte tenue à l'occasion de la 455ème réunion. Bruxelles, les mardi 6, mercredi 7 et jeudi 8 février 1968. CM2 1968/100, HAEU.
5. Anjo Harryvan, *In Pursuit of Influence the Netherlands' European Policy during the Formative Years of the European Union, 1952–1973* (Brussels: P.I.E. Peter Lang, 2009): 197.
6. Quoted in Antonio Varsori, *La Cenerentola d'Europa: l'Italia e l'integrazione europea dal 1946 ad oggi* (Soveria Mannelli, Catanzaro: Rubbettino, 2010): 210. The author's translation from Italian.
7. Allocution de M. Wilson au Diner du Guildhall, Londres, le 14 novembre 1967. DE-CE Coopération économique 735, AMAE.
8. Melissa Pine, "Perseverance in the Face of Rejection: Towards British Membership of the European Communities, November 1967–June 1970," In *Aufbruch zum Europa der zweiten Generation: die europäische Einigung 1969–1984*, eds. Matthias Schönwald and Franz Knipping (Trier: Wissenschaftlicher Verlag Trier, 2004), 292.
9. T.W.Garvey to Mr Hancock, May 14, 1968. PRO/FCO 55/48, TNA.
10. Official Committee on the Approach to Europe. Technology and industrial integration in Europe. Note by the Secretaries, March 28, 1968. PRO/FCO 55/48, TNA.
11. Susanna Schrafstetter and Stephen Twigge, "Spinning into Europe: Britain, West Germany and the Netherlands – Uranium Enrichment and

the Development of the Gas Centrifuge 1964–1970," *Contemporary European History* 2 (2002): 253–272.

12. Pine, "Perseverance," 297.
13. Dr. Luns and technology. Note of C. O'Neill to Mr. Hancock, February 1, 1968. PRO/FCO 55/59, TNA.
14. Note à l'attention de Monsieur le vice-president Hellwig: Réunion du groupe Politique de la recherche scientifique et technique (Groupe Maréchal) prévue pour les 15–16 février 1968. 375 Hellwig N/1359, BAK. The inconsistent policy of the Italians was perhaps due to Italian domestic issues, such as the strong opposition to integration of the left-wing parties and the problematic economic and social situation, which took the country's focus away from the European affairs. Maria Eleonora Guasconi, "Italy and the Hague Conference of December 1969," *Journal of European Integration History*, 9 (2003): 101–117.
15. EEC Ministerial Meeting on Technology. Note of Alan Smith, October, 25, 1967. PRO/FCO 55/59, TNA.
16. Premier Ministre. SGCI: Aide au développement industriel de la recherche. Aspects communautaires susceptibles d'être évoqués lors des conversations franco-allemandes: Réunion des 15 et 16 mars 1968, Paris, le 7 février 1968. 19850664, art. 2, ANF.
17. Comité des Représentants permanents. Projet de compte rendu sommaire de la réunion restreinte tenue à l'occasion de la 455ème réunion, Bruxelles, les 6, 7 et 8 février 1968. CM2 1968-100, HAEU.
18. Renvoi de la réunion du "Groupe Maréchal" prévue pour les 1 et 2 février 1968. Lettre de la Commission des Communautés européennes signée par M. Lionello Levi-Sandri, Vice-Président en date du 7 février 1968 à M. Maurice Couve de Murville, Président du Conseil des Communautés européennes. CM2 952, HAEU.
19. Comité des Représentants permanents. Projet de compte rendu sommaire de la réunion restreinte tenue à l'occasion de la 456ème réunion, Bruxelles, les 13, 14, 15 et 16 février 1968. CM2 1968-101, HAEU; Note à l'attention de MM. les membres de la Commission: Réunion du Comité des Représentants permanents en date du 15 février 1968. CEAB-3746/2, HAEU; Herrn Dr. Hellwig: Gruppe Maréchal, Brüssel, den 26. Februar 1968. 375 Hellwig N/1359, BAK.
20. Gespräch des Bundesministers Brandt mit dem Staatsminister im britischen Außenministerium, Lord Chalfont, den 8. Januar 1968. AAPD 1968, n°5.
21. WEU-Ratstagung am 30. Januar 1968: Wissenschaftlich-technologische Zusammenarbeit (Britischer Beitritt zu Euratom), Bonn, den 26. Januar1968. B 35-164, PA AA. The author's translation from German.

22. Note sur la coopération technologique européenne. Document remis par la délégation belge au Conseil des Ministres des Communautés, le 29 février 1968. CM2 1968-952, HAEU.

23. Deutsche Vorschläge für die handelspolitische und technologische Zusammenarbeit zwischen den Mitgliederstaaten der Europäischen Gemeinschaften und anderen europäischen Ländern, Brüssel, den 7 März 1968. CM2 052, HAEU. The Germans were also toying with the idea of easing the British entry by first inviting the UK to join Euratom—which was a strategy, they assumed, that the French would find hard to oppose. This option, however, was welcomed with little enthusiasm in London, and neither was it unanimously supported in the German administration. Beitritt Großbritanniens zu den europäischen Gemeinschaften; Beitritt zu Euratom, Bonn, den 8. Januar 1968. B 35-164, PA AA; Beitritt Großbritanniens zu Euratom, Bonn, den 15. Januar 1968. B 35-164, PA AA.

24. Extrait du procès-verbal de la réunion restreinte tenue à l'occasion de la 26ème session du Conseil, Bruxelles, le 9 mars 1968. CM2 1968-952, HAEU. This stand was repeated also in the following Council meeting of April 5, 1968. Extrait de procès-verbal de la réunion restreinte tenue à l'occasion de la 30ème session du Conseil, Luxemburg, le 5 avril 1968. CM2 1968-953, HAEU.

25. Ministerkonferenz technologische Zusammenarbeit im Anschluss an Drahtbericht nr. 149 vom 11. März 1968, Haag, den 12. März, 1968. B 35-164, PA AA.

26. Technologische Zusammenarbeit zwischen EWG und Großbritannien, Rom den 14. März 1968. B 35-164, PA AA.

27. Gespräch des Bundesministers Brandt mit dem französischen Außenminister Debré in Paris, den 7. September 1968, AADP 1968, n°287.

28. Ausführungen des französischen Wirtschaftsministers R. Galley vor der Nationalversammlung am 18./19. November 1968, Bonn, den 9. Dezember 1968. B 35-164, PA AA. This was stressed in the discussions between the French Foreign Minister Couve de Murville and the British Foreign Secretary Michael Steward in Paris on April 24, 1968. *Documents Diplomatiques Français* 1968, Tome I. Brussels: P.I.E. Peter Lang 2009, 702 (doc. 263) (hereafter DDF).

29. Especially the Belgian government felt uneasy about the impasse. Note de la délégation Belge sur la coopération technologique européenne, Bruxelles, le 29 mars 1968. CM2 1968-952, HAEU.

30. Extrait de procès-verbal de la réunion restreinte tenue à l'occasion de la 30ème session du Conseil, Luxemburg, le 5 avril 1968. CM2 1968-953,

HAEU; Poursuite des travaux en matière de coopération technologique, Strasbourg, le 15 mai 1968. CM2 1968-953, HAEU.

31. John W. Young, "Technological Cooperation in Wilson's Strategy for EEC Treaty," in *Harold Wilson and European Integration: Britain's Second Application to Join the EEC,* ed. Oliver J. Daddow (London: Frank Cass, 2003): 97–99. Besuch des britischen Außenministers in Bonn am 24. Mai 1968. Mit dem Beitrittsantrag Großbritanniens zusammenhängende Fragen: Technologische Zusammenarbeit, Bonn, den 21. Mai 1968. B 35-164, PA AA.

32. Extrait de procès-verbal de la réunion restreinte tenue à l'occasion de la 51ème session du Conseil, Bruxelles, les 4 et 5 novembre 1968. CM2 1968-954, HAEU; N. Piers Ludlow, *The European Community and the Crises of the 1960s: Negotiating the Gaullist Challenge* (London and New York: Routledge, 2006): 152–153.

33. 492. Tagung des Abschusses der Ständigen Vertreter, Brüssel, den 4. Dezember. B 35-164, PA AA.

34. Recherche scientifique et technique. Annex I au document I/19/. CM2 1968-954, HAEU.

35. Commission of the European Communities: *Second General Report on the Activities of the Communities 1968.* Brussels-Luxembourg, February 1969, 207.

36. Sujets à traiter dans le discours-programme du président Malfatti. Note de A. Spinelli, le 4 aout, 1970. AS-00270, HAEU.

37. Belgium, the Federal Republic of Germany, France, Great Britain, Italy and the Netherlands.

38. Christopher Layton, *European Advanced Technology. A Programme for Integration* (London: George Allen & Unwin, 1969): 161–164.

39. John Krige, *A History of the European Space Agency, 1958–1987* (Noordwijk: ESA, 2000): 344, 366–368.

40. In addition to the countries participating in ELDO, the members of ESRO included Denmark, Spain, Sweden, and Switzerland. Austria and Norway had observer status.

41. Krige, *A History of the European Space Agency,* 350–363.

42. Allocution du Dr. F. Hellwig, Vice-Président de la Commission des Communautés Européennes. Recherche-développement et concurrence dans les Communautés européennes. Actes du Colloque international organisé par le Centre de Documentation et de Recherches Européennes de Grenoble et la Commission pour l'Etude des Communautés Européennes, Grenoble, les 16 et 17 avril 1970s. AS-00319, HAEU. The author's translation from French.

43. Ibid.

44. Layton, *European Advanced Technology,* 228.

45. Ludlow, *The European Community*, 169–170.

46. Ibid., 152–153.

47. La Grande-Bretagne et l'accord de Bruxelles sur la technologie, Londres, le 11 décembre 1968. DE-CE Coopération économique 735, AMAE; La Haye, le 14 décembre 1968, DE-CE Coopération économique 735, AMAE. At the same time, France's readiness to compromise had its limits: a Council discussion on Euratom's new research program that took place less than two weeks after the meeting of December 10, exposed again the classic constellation in which the French categorically turned down the compromise proposals presented by the others. Any prospect of expanding the program to include new areas of research was resolutely rejected by Research Minister Robert Galley, who went as far as to state that Euratom, in the worst case, should be liquidated altogether. Tagung des Rates der EG am 20. und 21 Dezember 1968: Künftige Forschungstätigkeit Euratoms, Brüssel, den 22. Dezember 1968. B 35 166, PA AA.

48. Lord Chalfont's Visit to Dr. Luns 23 May (1968), speaking notes. PRO/FCO 55/61, TNA.

49. La Grande-Bretagne et l'accord de Bruxelles sur la technologie, Londres, le 11 décembre 1968. DE-CE Coopération économique 735, AMAE.

50. Ludlow, *The European Community*, 163–165.

51. 491. Tagung des Ausschusses der St. V. am 29.11.1968: Europäische technologische Zusammenarbeit, den 27. November 1968. B 35-164, PA AA. The author's translation from German.

52. Ibid., 151–152.

53. Margit Szöllösi-Janze, "Wissenschaft – ein neues Konzept zur Erschließung der deutsch-deutschen Zeitgeschichte," in *Koordinaten deutscher Geschichte in der Epoche des Ost-West-Konflikts*, ed. Hans Günter Hockerts (München: Oldenburg, 2004), 294.

54. Technologische Zusammenarbeit. Vorschläge und Argumente für Diskussion. B 35-164, PA AA. The author's translation from German.

55. "Technologische Lücke. Unbewältigte Zukunft," *Der Spiegel*, 9 (1969). The author's translation from German.

56. Helmuth Trischler, "Das bundesdeutsche Innovationssystem in den 'langen 70er Jahren': Antworten auf die 'amerikanische Herausforderung,'" in *Innovationskulturen und Fortschrittserwartungen im geteilten Deutschland*, eds. Johannes Abele, Gerhard Barkleit, and Thomas Hänseroth (Köln: Böhlau, 2001): 48–49. See also: Helmuth Trischler, "Die 'amerikanische Herausforderung' in den 'langen' siebziger Jahren: Konzeptionelle Überlegungen," in *Antworten auf die amerikanische Herausforderung. Forschung in der Bundesrepublik und der DDR in den 'langen' siebziger Jahren*, eds. Gerhard Ritter, Helmuth Trischler, and

Margit Szöllösi-Janze (Frankfurt and New York: Campus, 1999): 16; Johannes Abele, "Innovationen, Fortschritt und Geschichte. Zur Einführung," in *Innovationskulturen und Fortschrittserwartungen im geteilten Deutschland*, eds. Johannes Abele, Gerhard Barkleit, and Thomas Hänseroth (Köln: Böhlau, 2001): 16.

57. Hans-Willy Hohn, "'Big Science' als angewandte Grundlagenforschung. Probleme der informationstechnischen Großforschung im Innovationsystem der 'langen' siebziger Jahre," in *Antworten auf die amerikanische Herausforderung. Forschung in der Bundesrepublik und der DDR in den 'langen' siebziger Jahren*, eds. Gerhard A. Ritter, Margit Szöllösi-Janze, and Helmut Trischler (Frankfurt/Main: Campus-Verlag, 1999): 50–51.

58. The organizations officially belonging to UNICE were Bundesverband der Deutschen Industrie, Bundesvereinigung des Deutschen Arbeitgeberverbände, Fédération des Industries Belges, le Conseil Patronat Français, la Confederazione Generale dell'Industria Italiana, la Fédération des Industriels Luxembourgeois, le Verbond van Nederlandsche Werkgevers, la Centraal Social Werkgevers- Verbond, la Katholiek Verbond van Werkers-Vakvereinigingen and le Verbond van Protestants-Christelijke Werkgevers in Nederland. Luciano Segreto, "L'UNICE et la construction européenne (1947–1969)," in *Inside the European Community. Actors and Policies in the European Integration 1957–1972*, edited by Antonio Varsori (Baden-Baden: Nomos and Bruxelles: Bruylant, 2006), 201–202.

59. De H.M. Claessens à Jean Rey, le 16 juillet, 1968. BAC 118/1986-1393, HAEU.

60. UNICE: Possibilités d'une politique de la recherche dans la Communauté, le 19 juillet 1968. BAC 118/1986-1393, HAEU; UNICE: Mémorandum sur les activités de recherche des Communautés européennes, le 12 décembre 1968. CM2 1968-954, HAEU.

61. UNICE: Mémorandum sur les activités de recherche des Communautés Européennes, le 12 décembre 1968. CM2 1968-954, HAEU. The author's translation from French.

62. Western European Union was an international military alliance established by the Brussels treaty in 1948 and dissolved in 2011.

63. Aufzeichnung: Frage der Abgeordneten Herklotz (SPD) betr. die Empfehlung 174 der Versammlung der West-Europäischen Union von 18 Oktober 1968, Bonn, den 11. Dezember 1968. B 35-164, PA AA. The author's translation from German.

64. Technologische Zusammenarbeit zwischen den Europäischen Gemeinschaften, ihren Mitgliedstaaten und Grossbritannien, sowie

anderen beitrittswilligen Staaten, Haag den 10. September 1968. B 35-164, PA AA.

65. Communautés Européennes; Le Conseil: Résolution sur la politique européenne de la recherche et de la technologique adoptée par l'Assemblée lors de sa séance du 1ᵉʳ octobre 1968, Strasbourg, le 2 octobre 1968. 19850664, art. 2, ANF.

66. Ibid.

67. Commission of the European Communities: *Third General Report on the Activities of the Communities 1969* (Brussels-Luxembourg, February 1970), 233.

68. Sitzung der Gruppe Maréchal vom 7. und 8. Januar 1969. Memorandum an Dr. F. Hellwig von H. Michaelis, Bruxelles, le 8 janvier 1969. 375 Hellwig N/1359, BAK.

69. Entwurf eines Exposés über den Stand der Arbeiten der Gruppe Maréchal, Brüssel, den 12. Dezember 1969. 375 Hellwig N/1359, BAK; Extrait du procès-verbal de la 72e session du Conseil tenue à Luxemburg le 30 juin 1968. CM2 1969-1013, HAEU, Gesprächsvorschlag: Aigrain-Bericht, Bonn, den 18 Juli 1969. B 35-202, PA AA.

70. Aide-mémoire de la réunion du groupe de hauts fonctionnaires de la recherche scientifique tenue le 9 juillet 1969. CM2 1969-1016, HAEU.

71. Politik auf dem Gebiet der wissenschaftlichen und technischen Forschung: Inoffizielle Reaktionen der Drittstaaten. Brüssel, den 1. Juli 1969. B 35-202, PA AA.

72. Richard T. Griffiths, "A Dismal Decade? European Integration in the 1970s," In *Origins and Evolution of the European* Union, eds. Desmond Dinan (Oxford and New York: Oxford University Press, 2006), 172.

73. Note: Relevé des décisions prises par le Conseil lors de sa 84ème session tenue le 28 octobre 1969, Bruxelles, le 30. octobre 1969. CM2 1969-1014, HAEU; Commission of the European Communities, *Third General Report*, 222–223.

74. 575. Sitzung der Ständigen Vertreter, Brüssel, den 6. November 1970. B 35-225, PA AA; 576. Sitzung der Ständigen Vertreter, Brüssel, den 16. November 1969. B 35-225, PA AA; Comité des représentants permanents. Projet de compte rendu sommaire de la 579ᵉᵐᵉ réunion tenue à Bruxelles, les 2, 3 et 4.12.1970. CM2-105, HAEU.

75. Europäische Zusammenarbeit in Forschung und Entwicklung. Drittstaaten, die bisher ihr Interesse an Zusammenarbeit bekundet haben. Brüssel, den 20. Oktober 1969. B 35-202, PA AA; Note des conseillers commerciaux des pays de la C.E.E. en poste à Helsinki (novembre 1969). B 35-225, PA AA; 259. Tagugng der Ständiger Vertreter: Europäische Zusammenarbeit in Forschung und Entwicklung, Bonn, Oktober 1969. B 35-202, PA AA; Coopération scientifique et technique entre la

Communauté et la Yuogoslavie. Bruxelles, le 22 avril 1970. 19900634, art. 318, ANF.

76. Cold War logic mattered to the French decision makers, who advocated the invitation of Yugoslavia as a means of intensifying the relations between Belgrade and the EC (as well as those between Belgrade and Paris). For France, COST cooperation offered a good opportunity to capitalize on the Yugoslavian willingness to distance itself from Eastern technology and the socialist camp. Ibid.; Yougoslavie et Marche commun, Belgrade, le 30 avril 1970. 19900634, art. 318, ANF.

77. Commission des Communautés européennes: Note pour le Comité des Représentants permanents, Bruxelles, le 10 octobre 1969. B 35-202, PA AA.

78. Weisung: Wissenschaftliche und technische Forschung, Bonn, den 30. September 1969. B 35-202, PA AA; Sitzung der Ministerrats der Europäischen Gemeinschaften am 28. Oktober 1969 in Luxemburg: Europäische Zusammenarbeit auf dem Gebiet der wissenschaftlichen und technischen Forschung, Bonn, den 15. Oktober 1969. B 35-205, PA AA; Der Bundesminister für Bildung und Wissenschaft an das Sekretariat der Arbeitsgruppe der wissenschaftlichen und technischen Forschung, Bonn, den 3. Dezember 1969. B 35-202, PA AA; Mémorandum de la délégation allemande sur les problèmes de cohérence, Bruxelles, le 15 décembre 1969. CM2 1969 1015/a, HAEU, Europäische technologische Zusammenarbeit, Bonn, den 14. April 1970. B 35-225, PA AA.

79. Weitere deutsch-jugoslawische Zusammenarbeit bezüglich des jugoslawischen Verhältnisses zur EWG: Wissenschaftlich-technologische Zusammenarbeit, Bonn, den 9. Juni 1970. B 35-225, PA AA.

80. Aide-mémoire de la réunion du groupe de hauts fonctionnaires de la recherche scientifique tenue le 9 juillet 1969. CM2 1969-1016, HAEU.

81. Comité Interministériel pour les Questions de Coopération Economique Européenne, Paris, le 24 novembre 1971: Coopération Scientifique européenne. Conférence des Dix-neuf Ministres de la Science. Pompidou papers, AG / (2) /195, Archives Nationales Paris (hereafter ANP). The author's translation from French.

82. Luca Guzzetti, *A Brief History of European Union Research Policy* (Luxembourg: European Communities, 1995): 42.

83. Comité Interministériel pour les Questions de Coopération Economique Européenne, Paris, le 24 novembre 1971: Coopération Scientifique européenne. Conférence des Dix-neuf Ministres de la Science. Pompidou papers, AG / (2) /195, ANP.

84. Ibid.

85. Draft minutes of the first ministerial conference on European cooperation in the field of scientific and technical research, held at Brussels on Monday 22 and Tuesday 23 November. CM2 1971-1091.1, HAEU.

86. Young, "Technological Cooperation," 110.
87. Aigrain and the EEC entry. Note by C.J. Thomas to Mr. Auland, October 22, 1969, PRO/FCO 55/193, TNA.
88. Quoted in Aked and Gummett, "Science and Technology," 270–294.
89. Draft of a general resolution to be adopted by the conference. Annexe V to the draft minutes of the first ministerial conference on European cooperation in the field of scientific and technical research, held at Brussels on Monday 22 and Tuesday 23 November. CM2 1971-1091.1, HAEU.
90. Directorate General External Relations (I/A/2), Brussels, July 23, 1971. BAC 23/1979-830, Commission Historical Archives, Brussels (hereafter CAB).
91. Poursuite des travaux en matière de coopération technologique, Strasbourg, le 15 mai 1968. CM2 1968-953, HAEU.
92. Extrait de procès-verbal de la réunion restreinte tenue à l'occasion de la 30ème session du Conseil, Luxemburg, le 15 avril 1968. CM2 1968-953, HAEU.
93. Ludlow, *The European Community*, 184.
94. Pierre Aigrain to Pierre Harmel, Paris, July 18, 1970. CM2 1970-1094, HAEU.
95. To name an example, the German members of that committee all had their backgrounds in different ministries (Bundesministerium für Wissenschaftliche Forschung, Bundesministerium für Bildung und Wissenschaft and Bundesministerium für Wirtschaft). List of experts who took part in the work of the Committee of Senior Officials on Scientific and Technical Research. B 35-311, PA AA.
96. Ibid.
97. Draft minutes of the first ministerial conference on European cooperation in the field of scientific and technical research, held at Brussels on Monday 22 and Tuesday 23 November. CM2 1971-1091.1, HAEU.
98. Quoted in Laurence Jourdain, *Recherche scientifique et construction européenne. Enjeux et usages nationaux d'une politique communautaire* (Paris: L'Harmattan, 1995): 63. The author's translation from French.
99. Quoted in ibid., 62–63.
100. Fifteen nation group on technology. Note by C.L. Silver, November 24, 1970. PRO/FCO 55/481, TNA.
101. Requests for participation came also from some non-European countries, such as Israel and Canada, but on this point the consensus remained firm: COST would not be extended to other continents. Weitere deutsch-jugoslawische Zusammenarbeit bezüglich des jugoslawischen Verhältnisses zur EWG: Wissenschaftlich-technologische Zusammenarbeit, Bonn, den 9. Juni 1970. B 35-225, PA AA.

CHAPTER 7

Contrasting Visions and Continuing Struggle

The year 1969 was a decisive year for the Community. In April, French President Charles de Gaulle resigned. Although his successor Georges Pompidou was initially careful in his European policy, the regime change in Paris indicated that the French attitudes toward European integration, which during de Gaulle years had been characterized by nationalism and support for intergovernmental solutions, were changing and that a new start for the EC was possible. Most importantly, even before the election, Pompidou had shown signs of flexibility on the Community enlargement.[1] The struggle about British membership had blocked progress in important Community initiatives. In research, it had led to COST, which the supporters of an EC-centric vision for European collaboration regarded as a deviation from the planned course. Moreover, with the completion of the common market in summer 1968, the Community's transition period was over, and the time had come to look for new beginnings. The spirit of a new start was expressed by Ralf Dahrendorf in *Die Zeit* in July 1971:

> The Treaties of Rome and Paris have started a process of European integration that has achieved a lot. But this development has been exhausted today. …The impulses that deliver what the European States of today and tomorrow, the Member States of the European Community and the newcomers of 1973, expect, are no longer produced by the First Europe. Instead, there must be a new beginning next to the remnants of the old achievements.[2]

© The Author(s) 2020
V. Mitzner, *European Union Research Policy*, Europe in Transition:
The NYU European Studies Series,
https://doi.org/10.1007/978-3-030-41395-8_7

Essentially, the new departure necessitated new tasks. The "Second Europe" would not just be a common market with an agricultural policy (as the "First Europe" had been) but a broader, more versatile Community engaged in a number of policy fields.

These changes dramatically altered the conditions in which research policy had been discussed in the previous years. With the renewed thrust of the Community project, research policy was recognized as one of the key components of the "Second Europe," while COST, as an intergovernmental arrangement principally designed to ease the British entry, appeared increasingly obsolete. Here, continuity was important. The fact that research had already been discussed within the EC, and that it was widely regarded as a vital dimension of economic policy, considerably improved its chances of inclusion in the Community's future agenda. Despite the rising criticism of growth-oriented research policy in many Western European countries, science retained its high societal status and was further harnessed for achieving even greater economic expansion.

This chapter shows how the Community-centric concept of research policy returned as a corollary of the vision of a "Second Europe," and how the strong ideational continuity in European research policy, supported by the Commission's ability to answer to the rising ideational challenges, shifted the EC debate on research. But it also demonstrates the difficulty of creating a common policy for research. An excellent example of this prevailing challenge was the establishment of the European Science Foundation (ESF), which despite the Commission's efforts to realize it within the EC, came into being as a non-Community enterprise led by powerful national science administrations, anxious to preserve their independence and skeptical of proposals that could lead to increasing political control.

A symbolic event in the discussion on second departure for the Community was the summit of The Hague, which President Pompidou proposed in summer 1969. The original idea of the French leader was to organize a gathering where Europe's heads of state and the government could discuss the triptych of *achèvement* (completion of the Community's initial agenda), *approfondissement* (deepening cooperation into new areas), and *élargement* (widening the EC's membership). In Pompidou's view, the enlargement of the Community could only be made possible if all of the tasks that were intended to be finished during the transitional period were first completed, and if the priorities of the EC's future development were set. In that case alone, the Six could begin membership negotiations well prepared and without running the risk of jeopardizing either the earlier achievements of the Community or the prospects for its future.[3]

The Hague meeting of December 1 and 2, 1969, marked a moment where the standstill of the Community was finally overcome.[4] The event witnessed a strong willingness to press ahead with integration and a determination to solve the disputes that had been hampering the development in the preceding years. After the conference, there was no doubt that the French veto on the enlargement had been lifted and membership talks would begin by the following summer. But progress was also made in other fields: the EC became its own financial resource, and the participating countries reached an agreement on the definitive regime of the common agricultural policy. In addition, new areas of the Community's activity, such as monetary policy, social policy, and political and technological cooperation, were considered.[5]

Despite its prominent role in the earlier discussions over the enlargement, science and technology were not at the center of the debates at The Hague. However, for the advocates of a common research policy, the summit carried special importance. First, resolving the struggle over the enlargement allowed the Community to move on and to discuss its future agenda that would eventually include research.[6] Second, science and technology were included in paragraph 9 of the final declaration of the meeting, stating that "[w]ith regard to the technological activity of the Community, they [the Heads of State and Government] have reaffirmed their desire to pursue more intensively the activity of the Community aiming to coordinate and encourage research and industrial development in the main high-tech sectors, in particular through Community programs, and to provide the financial resources for that purpose."[7] It is interesting that the paragraph mentioned Community programs while including no reference to the Aigrain group proposals for the intergovernmental model of COST cooperation.

For those in favor of a common research policy, Article 9 of The Hague communiqué offered a tangible reference point in the continuing discussion about the EC's role in the field. In particular in the Commission corridors, The Hague declaration was interpreted as a valid mandate for the creation of a research policy that would also contain supranational elements.[8] This was despite the fact that, actually, there was no direct reference to a general research policy in the summit communiqué. Nevertheless, in June 1970, the Commission proposed a number of activities in the main sectors of research cooperation and suggested that the six member states should initiate prior consultations on all major R&D projects. This would be important for obtaining an overall picture of research policy in the Community as well as for intensifying cooperation between the member states. Likewise, the Commission suggested establishing new institutional structures providing "a permanent means" of preparing and carrying out "comprehensive programs." These would include a European Research and Development Committee (ERDC) responsible for preparing the decisions taken by the Community authorities and a European R&D Agency (ERDA) responsible for implementing the executive action worked out by the ERDC. The ERDA would also administer the funds made available to the Community.[9]

Commissioner Altiero Spinelli and his team developed these ideas further in a new proposal that they handed to the Council in November.[10] There, the ERDC was envisaged as having the following tasks: to define the domains or sectors of the Community activities, to elaborate common programs, to define the objectives and forms of cooperation of the Community with third countries and international organizations, and to define and propose the means of intervention or execution to realize the agreed objectives, including the organization of information centers, the harmonization of public initiatives, and the allocation of finances for certain R&D programs. The ERDC would consist of high officials responsible for R&D in the member states, representatives of the institutions promoting research and persons from the universities, industry, and labor unions.[11] Although the Commission emphasized R&D, it was clear that the new Community activities would also include fundamental research, which was the "source of technological development." In fact, the second Spinelli paper presented five distinct categories of action: pure fundamental or oriented research, applied research, public services, industrial development, and the environment. Activities in the first category would include the promotion of the cooperation between different groups, the mobility of researchers, teachers, and students, and the circulation of information between the countries. In addition, at times, "a more direct intervention at a European level" would be necessary.[12]

The Commission's rationale behind these new institutions was that the existing structures for Community R&D were insufficient for responding to The Hague resolution's global mandate.[13] Particularly disappointing was COST cooperation. In February 1971, the Brussels executive listed the shortcomings of this undesired initiative: in most cases, the activity concerned only modest pilot projects that could not respond to the fundamental needs of European scientific and technological cooperation. Furthermore, out of the 47 projects initially considered by the member states, less than a half had finally been adopted. Lastly, the geographically wide membership basis of COST did not give much indication of a rapid and efficient achievement of the works that the EC had started in the field.[14] Equally as bad, COST symbolized a type of intergovernmentalism that the Commission least wanted to support in Europe. From the Brussels point of view, intergovernmentalism came at the expense of its own status in the EC; any new mode of dealing between the governments would signify loss of supranationalism and marginalization of the Commission. In addition, all this was about to happen at a moment when the first phase of integration had been concluded and the Community was looking for new tasks. N. Piers Ludlow has argued that in the long run, maintaining and periodically fiddling with the existing *acquis communautaire* would hardly have justified the Commission bureaucracy in its actual size. Before long, the Commission would need new assignments—and a science and technology policy could eventually provide them. Given its specific, though increasingly ambiguous, allure in the eyes of the wider public, research would be well suited to underlining the merits of European unification. Moreover, it offered a terrain in which the Commission could relatively easily establish expertise useful for consolidating its political position.[15] For all that, it was necessary to oversee that the new activities remained within the EC framework where the Commission had better chances of preserving a central role.

The divergence of opinion between the Commission and the EC member governments on these matters manifested itself at the Council meeting of December 16 and 17, 1970, which discussed the Spinelli papers. Most enthusiastically the Commission's proposals welcomed the Belgian Research Minister Théo Lefèvre, who declared that an objective of the Community should be better organization and greater coherence of the activities in the domain of scientific and technological research.[16] This unsurprising position was in line with the continued support of the Belgian government for the initiatives of a common research policy.[17] The Belgians agreed with Spinelli that COST was not compatible with the objectives of the Community as they were expressed at The Hague conference.[18] On the opposite ends were the French: Paris could only envision Community activity if it was limited to the domains that already had their place in the Treaties. This marked a clear difference to the Spinelli proposals, which the French Permanent Representation in Brussels interpreted as nothing less than an attempt by the Commission to dominate all research-related activities of the member states.[19]

But French policy was not entirely rigid or without ambivalence. After the resignation of President Charles de Gaulle in 1969, there had been some indications of a renewed French interest in advocating the Community's activities in research. This went together with the more general reorientation of France's European policy, characterized by greater openness toward Community initiatives. Surprisingly perhaps, it was President Pompidou who was behind the introduction of the paragraph 9 on research of The Hague communiqué. Although Pompidou's ulterior statements suggested that the President only meant cooperation in certain well-defined projects,[20] it is clear that France wanted to keep the issue on the agenda. Moreover, in March 1970, the French Research Minister Maurice Schuman submitted a memorandum to the EC Council that not only called for reinforcing European industrial structures, but also included proposals for joint activity in some high-technology sectors.[21] The document was so different from France's earlier statements that the German Foreign Ministry even saw there "[a] fundamental reorientation of French technology policy within the Community."[22]

In Paris it was well understood that joint efforts in scientific research could offer benefits to France. Moreover, after the financial crisis of May 1968, France was no longer in the leading position in the coordination of economic and monetary policies within the Community,[23] and Elysée was aware of its limited room of maneuver in the other Community affairs. Self-assertiveness and nationalist arrogance were likely to turn out to be counterproductive and harm France's own interests. In addition, some international developments contributed to reconsideration of the French approach to the proposals for a common research policy: in February 1971, the United States unilaterally renounced Aerosat, an aviation navigation satellite system, which had been planned within NASA and ESRO. The cancellation of this transatlantic space venture especially angered the French who interpreted the incident as a clear sign of American disinterest in European affairs. "For the French," John Walsh wrote in *Science* in the following month, "the Aerosat incident seems to have crystallized the feeling that, in matters of technological, the Americans will act from now on only when they have calculated the net gain or loss for the United States."[24] This observation, which effectively reinforced the already existing French impression of the Americans as untrustworthy partners,[25] made it seem even more reasonable to focus on cooperation within a European framework.

But France's role as an initiator and innovator proved inconsistent. The French remained uncompromising in their rejection of supranationality. Laurent Warlouzet has also argued that in 1968 and 1969, the French project of European industrial policy, to which the initiatives of research policy were closely connected, lost its impetus. This was an immediate consequence of the departure of some *fonctionnaires modernisateurs* (officials keen to modernize France's policy) who had advocated a new industrial approach in the otherwise traditionalist circles of the French administration. After 1968, the French policy was increasingly focused on the theme of economic union, largely limited to the objectives of abolishing technical obstacles to exchanges and promoting certain elements of juridical and fiscal harmonization. The incentive measures for reinforcing European industrial structures, such as research policy, were delegated to a second plan.[26] This was also evident in a memorandum that the government produced in March 1970. "The French Memorandum," the Commission observed in April, "is mainly focused on industrial problems and that little is happening in terms of scientific research."[27]

In order to understand these ambiguities of French European policy, it is important to remember that the borderlines between the Community and the national spheres of political activity were not always strictly drawn. Many officials in Brussels had their background in national administration, and they eventually went back to their home countries after having left the Community institutions. For instance, François-Xavier Ortoli, director general of the DG III (Industry) 1958–1961, returned to France where he served as general-secretary of the Comité interministériel pour les questions de coopération économique européenne (SGCI) (1961–1964), technical councilor and then director in the Prime Minister Pompidou's cabinet (1962–1966), commissaire général au Plan (1966–1967), and finally held various ministerial portfolios between 1967 and 1971, including industrial and scientific development (1969–1972). In 1972 Ortoli, however, was back in Brussels as the president of the European Commission. During all these years, Ortoli had to defend many different positions. Ortoli is a perfect example of the fluidity of the borders between the different political realms and the continuous crossings and transit that created multiple loyalties and influenced policies on all sides.

France was not the only country who gauged the Commission's new initiatives on research with some caution. The Netherlands also wanted to proceed slowly on the issue. In addition to doubting the feasibility of the institutional solutions the Commission proposed, and supporting the French view of first proceeding with the definition of the content of the cooperation before considering institutional matters, the Dutch expressed reservations concerning the possibility of realizing a global policy in the short term.[28] For Britain, on the other hand, the ideal form of Community activity in the field of research continued to be *à la carte* cooperation. But as long as the membership negotiations were not concluded, London was forced to stay on a careful course. In July 1972, a Foreign Office note gave the following instructions: "I hope we shall react more positively, partly because, when we are outside the Community and the Commission has been to some trouble to keep in touch with us on these questions, we should not oppose flatly, unless we must, proposals to which the Commission attach importance; but more particularly because some aspects of these proposals are in our interest."[29]

Regardless of the multiplicity of the views of the issue, the discussions on the Spinelli proposals had demonstrated that the Community was willing to go beyond the strictly intergovernmental model that was emerging with COST. Moreover, the EC was widely recognized—even by the French—as the unequivocal center of the contemplated actions.[30] Concluding that the nuances of opinion separating the member states did not seem fundamental and that the creation of a real common policy within eight or ten years would still be possible, the Brussels institution felt encouraged to proceed with its project of elaborating the new administrative framework already outlined in the earlier proposals.[31]

In order to succeed with its plans, it was important for the Commission to create links with high-level officials responsible for science policy in the member states. Unofficial contacts were sought with a number of persons from national science administrations who were deemed to be in a position to influence diplomats and ministers.[32] But the attempts of the Commission to create these links did not mean the Brussels institution could have steered or manipulated the works being done in the expert groups and committees, especially those of the Council. This is clear in the Commission's relations to the ad hoc group studying the Commission's proposals of autumn 1970. In fact, the Commission received the report with disappointment. Irritation was stirred not only by the superficial fashion in which its documents were treated but also by the fact that the group, overstepping its mandate, questioned the readiness of the member states to realize a common research policy. The latter point particularly confused the Commission, thinking that with the declaration of the 1967 Luxembourg Council and in particular that of The Hague of December 1 and 2, 1969, the member states had already agreed to broaden the EC's activity in research.[33]

Another strategy that the Commission adopted to win supporters for its initiatives was stressing that the new economic austerity was no obstacle to giving the Community a more ambitious role in research and technology. The limited resources available for research funding at a national level could actually be a good reason for more far-reaching European cooperation. Rationalization, and effective division of labor at a European level, could produce savings and efficiency. In April 1971, Spinelli told the European Parliament:

> [T]he era of spontaneous technological development is coming to an end. In the course of last twenty years large industrial states have supported research projects on the basis of circumstances, on the hypothesis that all new technologies, especially spectacular ones, deserve support. Now we realize that the richest country in the world has started to question the rationale of this policy. …If the USA realizes that it cannot simultaneously conquer space, build a supersonic aircraft, improve conditions of life in its large conurbations and preserve the environment, then obviously Europe should become equally exacting in its choice of projects, eliminating duplication and as far as possible avoiding spending ten or fifteen years on what others have done before them.[34]

The Commission also responded to the new ideas that came with the discussion of the limits of growth. As often before, the Brussels body was inspired by the works in the OECD, which it had been following closely. Early in February 1970, the Commission representative at the OECD reported to his colleagues in Brussels that "[t]he OECD, which in the past had alerted member governments and public opinion to the existence of the 'technology gap' and which was the first to launch the idea that governments should have a science policy, is seriously questioning this notion."[35] A powerful advocate of revisionist views in Brussels was Sicco Mansholt, the European Commissioner for Agriculture and later the president of the European Commission. As chair of the Mansholt-Commission, set up in December 1971 by the three Dutch leftist opposition parties to study the issue of the "limits," Mansholt had an important role in the Dutch discussion on growth. The report of the Mansholt Commission was published a few months before the influential Club of Rome report, and it not only called for qualitative growth but also proposed shifting the priority from economic expansion to expansion of wealth that essentially would be reached by environmental protection and the reduction of industrial production. Moreover, the report envisaged Europe as a pioneer in the promotion of the concept of sustainable growth.[36] Crucially, Mansholt, who was a member of the Club of Rome and thus directly inspired by its debates,[37] did not restrict his activity to Dutch politics. In February 1972, he sent an open letter to the president of the European Commission, Franco Malfatti. Describing the letter as his "testament," Mansholt insisted on a new economic system based on strictly centralized planning and oriented toward preservation of ecological balance. In a Commission meeting in March 1972, Mansholt went as far as to suggest that the last year of his Commission should be devoted to a study of the radical political reorientation imposed by *The Limits of Growth* report. Although in Brussels Mansholt's ideas did not meet with much enthusiasm, they nevertheless increased the awareness of the environmental challenges of economic growth.[38]

In fact, the Commission modified its political vocabulary and moved to stress the qualitative aspects of growth. The need for a broader picture of growth was also emphasized in the preparatory documents of The Hague summit that highlighted the need for Community action on environmental protection. In the early 1970s, the Commission started to examine the possibilities of harmonizing environmental protection legislation in the EC member states and eventually to define a common environmental policy.[39] It is interesting that the creation of the Community's environmental policy followed a pattern very similar to that of research policy. First of all, in this context the EC was also dealing with a rather new concept: in the early 1970s, the notion of environment as an area of policy-making was only emerging. Moreover, the idea of environmental policy did not originate from the EC but the discourses of experts meeting in the framework of various international organizations where the role of the United States as a model and international agenda setter proved critical. Jan-Henrik Meyer has shown how the different EC institutions reacted to the international debate on environment and made it a part of their own political agenda.[40] Institutional innovations quickly followed. The establishment in 1971, upon the initiative of Altiero Spinelli, of an Environmental Service within the Directorate General of Industrial Affairs, anticipated the drafting of the first EC Environmental Action Program, scheduled for the period 1973–1977. Environment was also present in the Spinelli proposals that singled out environment as an important area of future activity.[41]

Creating a common environmental policy was not simple, however. The lack of a legal basis ultimately led to the subordination of the plans for economic objectives. Only in this way could the general Treaty Article 235 justify activities in the field.[42] In this regard, the formulation of the Commission's 1973 Environmental Action Program is very interesting. The scheme explicitly presented environmental policy as a contributor to "progress," generally understood as equivalent to economic growth: the goal of the new EC policy was to "reconcile expansion with the increasingly imperative need to preserve the natural environment." As Jan-Henrik Meyer has noted, "[s]uch a conceptualization had very little in common with the contemporary concern about the 'limits of growth' or with the policy of restraining resource use, advocated e.g. by Commissioner Mansholt."[43]

This was also the fate of the Commission's ideas for an overall research policy that would not necessarily be restricted to economic reasoning. In the end, the Commission had no other option than to present its proposals in an economic context where research had better chances to be seen as falling within the competences of the Community.[44] Somewhat contradicting itself, the Commission repeatedly underlined that research policy was not an end in itself but rather an element of the general economic policy aiming for growth.[45] Commissioner Spinelli even tried to address the juridical dilemma with his initiative to organize the General Directorates for which he was responsible. Beyond allowing a better overview of the Community's activities in research,[46] the explicit aim of the reform was to bring research closer to industrial policy.[47] Of course, this was intended as a temporary solution only. Sooner or later, the Treaty revision would give the Community proper competences in the field of research.[48]

Indeed, the Commission already started to draft blueprints for an eventual article on research. In November 1971, Christopher Layton, in an internal Commission note, suggested that the mandate for an industrial and technology policy should be similar to that of agricultural policy. He proposed the following formulation:

> To achieve a high rate of economic growth, a more rational use of Europe's resources in science and technology, and a stronger industrial base for Europe to meet the new needs of advanced societies, and to practice independent foreign policies, the Heads of Government are determined to

develop common policies in the field of industry, science and technology within the framework of the Community.

They are convinced that, while the execution of industrial research and development programs must continue to be carried out by a large number of organizations, ranging from private companies and national laboratories to independent European agencies, it is necessary to engage in a joint action to remove barriers to industrial integration, to pool resources for research and development in areas of common interest, and to identify common priorities and objectives as well as the means to achieve them.[49]

Layton's proposal is interesting in that it involves elements expressing both ideational continuity and discontinuity in research policy. While explicitly mentioning the objective of economic growth and the worries about European political independence, it also paid attention to the "new social needs" and the challenges emanating from limited resources as well as the questioning the objectives of technological development:

> Heads of Government recognize that new social needs and their increasing pressure on scarce resources require continuous reflection at the Community level on the priorities, goals and means to be used. The mechanism for choosing priorities for Europe is of particular importance at a time when scientific and technological potentials are constantly progressing and where, however, the means are limited while the goals of technological development are increasingly questioned.[50]

The Commission also had an additional reason to stick to the concept of research as an economic resource and growth as its major objective. To an important degree, the legitimacy of the Community and its institutions was (and is) built on economic success. In the 1960s, preserving peace in Europe started to lose its importance as the primary raison d'être of the Community project. Increasingly, it was wealth and prosperity that gave legitimacy to the EC. Continuous support of integration in the member states thus greatly depended on the economic benefit that came from belonging to the Community. Whatever new intellectual trends emerged from the OECD and other international forums, there was a rationale for the Commission to keep emphasizing growth and means that could lead to it.

The Commission's hard work to make the member states align on Community research policy was rewarded at the Summit of the Heads of State and Government, organized in Paris on October 19 and 20, 1972.

The final declaration of the meeting not only fixed a deadline for measures to be taken to advance efforts but also put even more emphasis on the Community-element of the common activities in research than the communiqué adopted at The Hague Summit three years earlier. The article on research and technology recalled that "objectives will need to be defined and the development of a common policy in the field of science and technology ensured. This policy will require the coordination, within the institutions of the Community, of national policies and joint implementation of projects of interest to the Community."[51] Contrasting with the declaration of The Hague that only mentioned common projects and the necessity of coordinating and encouraging industrial research, it was now about a "common policy" in which the EC institutions would play a crucial role.[52]

The summit's long list of achievements also included the establishment of a European monetary cooperation, an agreement on a common commercial policy toward Eastern Europe and on EC policy for the next round of General Agreement on Tariffs and Trade (GATT) talks, as well as the creation of a regional development fund. Last but not least, the national leaders "set themselves the major objective of transforming, before the end of the present decade... the whole complex of the relations of [the] member states into a European Union."[53] However, the success of the Paris meeting was ephemeral, and it rested on unstable foundations. In the early 1970s, tensions within the Community were growing again. Major institutional and policy differences, the cooling of relations between France and Germany, and finally, the gradual decline of the Commission's position all contributed to an atmosphere that hardly resembled the heady mood of The Hague.[54] A cable from the French permanent representation in Brussels only a few days before the conference captured the atmosphere: "It is a common complaint in these circles [activists of the European movement] that the next conference in Paris seems to be opening in a climate of disenchantment. We notice that the heart is no longer there and that the charm seems to be broken."[55] Despite the summit's impressive list of results, the prospects of a rapid political progress had dramatically diminished. As Derek Urwin has noted, "the final communiqué of the Paris summit was a political construction that could be accepted by everyone. It was perhaps a rhetorical statement first and foremost, with very little in the way of real commitment."[56] And this also applied to research policy.

Fundamentally, there was no significant change in the national positions on common research policy. The French had not renounced their basic opposition to the principle of an overall research policy as especially the Commission promoted it. Throughout the year, France had argued that no article of the EEC Treaty conferred general competences to the Community in research,[57] and predictably, both the British and French soon raised issues concerning the interpretation of the Paris Summit communiqué. While the British wanted to avoid too strict an interpretation of the concept of community action and thought that there could not be an overall research policy in the abstract but only common research policies in various sectors of the Community, the French argued that it had not been the intention of the Heads of State and Government to move toward a general delegation of powers to the Community institutions. The French underlined that the Paris conference merely paved the way toward implementing certain projects on a community level, and that the member states retained full responsibility for their policies.[58]

Unfortunately for the prompters of community-centric vision, in 1972, there was little hope in the German government playing its traditional role of arbitrator between the different positions. With the progress in the enlargement negotiations, for Bonn, research policy had lost its significance as a channel easing the accession of the applicant countries to the Community. Moreover, now there were also other, more urgent issues on plate. Compared to the preparations for British membership and the topical questions of monetary cooperation, research appeared a minor thing. A change in the internal power relations of the German government further strengthened this perception: before, the Foreign Ministry, which promoted a rather favorable attitude to the Community initiatives, had been able to count on the support of the Research Ministry, while the position of the Finance Ministry had been much more reserved. The appointment of Klaus von Dohnanyi as the German minister of research in March 1972, however, sharpened tensions within the German administration and effectively reduced Bonn's willingness to assume political leadership in this contentious issue.[59] Dohnanyi had close links to the Finance Ministry, where he had served as a state secretary for the Finance Minister Karl Schiller, and as a research minister, he advocated views that clashed with his ministry's earlier positions.[60]

While German engagement with the EC research policy debate became increasingly ambivalent, important changes took place also within the Commission. With the redistribution of the portfolios in January 1973, the responsibility for research, science, and education and the Joint Research Center (DG XII) went from Altiero Spinelli to Ralf Dahrendorf, whose approach differed from that of his federalist predecessor. A university professor with an academic career in both Germany and Great Britain, Dahrendorf had joined the Commission in 1970, first with the portfolio of external relations. In January 1973 that position went to Christopher Soames, and Dahrendorf was left with research, or as the Commissioner himself has put it, "Euratom things plus education–whatever that meant."[61] Dahrendorf, who not only knew the academic world but also was very familiar with intergovernmental tensions, was by no means naive about the political possibilities of the Community, and he approached his new position with a decent amount of pragmatism.[62]

In 1973, both Dahrendorf and Spinelli, who meanwhile had been appointed to run DG III (Industry), published work programs according to the stipulation of the final declaration of the Paris Summit. Spinelli's program, handed to the Council in May, called for elimination of technical barriers to exchanges, openness of public markets, and promotion of European enterprises, facilitating the circulation of information.[63] Dahrendorf's program was presented in autumn and envisaged the creation of a "European Research Area" that would involve coordinating national policies, promoting fundamental research, and ameliorating information management and technological forecasting. The Commissioner also proposed supportive activities in other policy sectors, such as agriculture, industry, energy, and social and development policies. In addition, he suggested the creation of a new Community organ, Comité de la recherche scientifique and technique (CREST), which would replace the PREST[64] and be charged with coordinating national policies with competences in all aspects of Community research.[65]

While taking some distance to industrial policy, which now was discussed within a separate DG and formulated in its own program, Dahrendorf's work program nevertheless mentioned growth: "Economic growth, which remains an important goal, makes it necessary to raise the level of research and education." There was no question of denouncing the objective of growth or the role of science in the pursuit of that objective. But Dahrendorf did not advocate growth for its own sake: "Growth must be put at the service of people and the improvement of the quality of life. In this respect, science and education must occupy a special place."[66] This approach reflects a broader discursive change in the Commission, and the revision of attitudes toward growth and science in Europe more generally.

Furthermore, if the EC's juridical limits had driven Spinelli to tie research closer to industrial and economic policies, Dahrendorf gave surprisingly little weight to those constraints.[67] The new research commissioner did not want the Treaties restrain his room for maneuver, and he indeed felt empowered to move to new territories. One of these novelties was the initiative of a European Research Area. The idea of coordinating the member states' national policies had figured in a number of earlier Community proposals, but it was only now that the concept was efficiently formulated and given an explicit name and some concrete content. Indeed, Dahrendorf's vision seems very familiar to the ERA that the EU has been trying to realize since 2000.[68] According to the 1973 work program, "[t]he European Community could and ought to make its contribution to overcome the limits of national thinking in the development of science and to create an effective single area for European science in which cooperation and competition complement each other in a sensible way." The objectives of such an area would be to

> facilitate the mobility of researchers in the Community; facilitate international meeting in the Community; stimulate European cooperation by concerted actions and projects; finding laboratories qualified for developing into special areas of research on a European level [and] set-up a "professional network" of such laboratories; co-ordination of costly long-term projects; common use of expensive instruments.[69]

The Commissioner's idea of a European Research Area, however, did not evoke much political interest. When the program was discussed within the COREPER, the focus was on the institutional aspects of the document, not on the research area.[70] The final compromise on CREST and the Community's action program that was adopted without modifications by Council on January 14, 1974,[71] nevertheless shows the continuing support for common research policy. CREST was to play a triple role in counseling the Community institutions, coordinating the national policies of the member states, and consulting activities with the third countries. The organ would be composed of high national officials and the representatives of the Commission.[72]

The Council meeting of January 14, 1974, also adopted a declaration regarding the European Science Foundation, a new body of European research cooperation established outside the Community structures. Since the early 1960s, the idea of a European Science Foundation had been floating around in different European forums,[73] but it was only in the early 1970s that the plans of establishing the foundation started to crystallize. The obvious reason for this was the rapid increase of informal collaborations between European academies and research councils: the formation of an informal association of ten European medical research councils in May 1971 was followed in February 1972 by the creation of the "Aarhus group" as a result of an initiative of the Nordic research councils and the Nederlandse organisatie voor zuiver-wetenschappelijk onderzoek.[74] This development soon prompted the Council of Europe[75] and the OECD[76] to consider the possibility of a new European body to promote fundamental research. But there was also another important driver behind the foundation initiative: in the early 1970s, basic research in many Western European countries was hit by government budget cuts imposed by the recession. Attempting to come into grips with the hardening economic situation and to find alternative funding, science administrators saw the foundation as an attractive way to access new resources.[77] These individuals were ready to see the emergence of an institution with important resources[78]—that could balance out diminishing national research budgets.

A crucial occasion in the foundation project was the third Parliamentary and Scientific Conference organized by the Council of Europe in Lausanne in April 1972. The conclusions of the conference advocated the setting up of a European Science Foundation for the purpose of developing and organizing scientific cooperation in various fields.[79] A European foundation for science was also included in the proposals that Commissioner Spinelli presented to the Council of Ministers in June 1972. Spinelli envisaged the creation of an independent body that would, through coordination, support the existing centers and associations. Spinelli's ESF would have had its own funds from the Community budget and from the contributions of public or private institutions in the member and non-member states participating in its work.[80]

The initiative was discussed by leading scientists from 13 European countries at an informal conference the following December. It is no coincidence that the conference, which was held by the Royal Society of London, took place in Britain: the British Foreign Ministry had been considering the idea of a "European fund for Science and Technology, financed by a levy on purchases from outside the community" already in 1967.[81] Crucially, in London, the foundation plans took a radically different course from that envisaged by the Commission. Spinelli's idea of organizing the foundation on the basis of the EC countries evoked strong reservations.[82] A Community or an intergovernmental arrangement was not seen as ideal because it could run the risk of being dominated by politicians and bureaucrats instead of scientists themselves. A more desirable way to promote European scientific cooperation would be the joint activity run by national academies and research councils across the Western part of the continent. The conclusion of the meeting was that steps should be taken to set up a common forum with an important amount of autonomy and with the view of organizing cooperation in a widest possible Western European framework.[83]

The initiative was further discussed in two subsequent conferences: in Munich in April 1973, in a conference organized by the Max-Planck-Gesellschaft; and in Gif-sur-Yvette in September 1973, in a meeting held by the CNRS. In Munich, the majority of participants, which consisted representatives of academies and research councils from ten European countries[84] as well as the representatives of the European Commission,[85] agreed to work toward the establishment of a European Science Foundation.[86] At Gif-sur-Yvette, the number of countries represented had grown to fifteen,[87] and a decision was reached about the foundation's creation. The foundation would be a nongovernmental organization with the objective of advancing cooperation in basic research, promoting mobility of researchers, assisting the free flow of ideas and information, and facilitating the harmonization of the basic research activities supported by its members.[88]

Although involved in the preparatory work, the European Commission soon found itself marginalized. The further the discussion about the foundation progressed, the more insecure the position of the Community became vis-à-vis this emerging collaborative framework. A preparatory paper for the Munich meeting reveals the perception of the leading European scientists and science administrators of the European Community: for them, the EC appeared to be a limited framework that was not in a position to speak on behalf of "Europe." Likewise, the architects of the ESF disliked the Brussels bureaucracy that they feared would restrain the freedom of scientific activity. On the other hand, the Community was recognized as a useful actor that, if it was not given a dominant role, could nevertheless make an important contribution.[89] In a disappointed tone Commissioner Dahrendorf's General Director Günter Schuster, who had been present at a preparatory group meeting in July

1973, remarked that none of the other participants had deemed it necessary to speak in favor of the Commission.[90]

Bearing in mind the Commission's disappointment with COST, it is surprising that the Brussels executive appeared fairly comfortable with the idea of establishing the ESF as a non-Community institution with a broader geographical basis. More important than having the Community label on the new foundation was the aim of securing its presence in the endeavor and ascertaining that, otherwise, the initiative was developing in the desired direction. This shows the new pragmatism that characterized Commissioner Dahrendorf's policy: if the Brussels institution could not push its own design through, then it should at least try to maximize its influence inside the foundation. Günter Schuster warned the Commissioner in May 1973 that[91] "Without the Community's participation approved by the Council of Ministers, an EFSF [ESF] would strengthen the idea of intergovernmental cooperation and weaken the idea of a more and more cohesive Community."[92] As a minimal solution, Schuster envisaged a financial contribution to the running costs of the ESF and to the foundation's projects that could be of interest to the EC. Permanent representation in the foundation's steering committee or a corresponding organ and the secretariat would also be necessary.[93]

Dahrendorf and Schuster probably also realized that compromising the relations with national research councils could make any future cooperation difficult if not impossible. By approving a different concept of the ESF, the Brussels institution avoided a break with its crucial partners and an otherwise inevitable and embarrassing political failure. Dahrendorf's and Schuster's academic background certainly helped to understand where the foundation's proponents, many of which the Commissioner knew personally, were coming from and what the Commission could realistically expect.[94] Furthermore, the objectives of the proposals from the national research councils and academies were not very different from those of the Commission: in both cases the core idea was to set up a relatively independent institution promoting fundamental research in Europe. A Commission note from May 1973 explicitly stated that the goals set by the Commission for the foundation were "practically identical" to the aims of the science councils and academies.[95]

Indeed, in Munich, Dahrendorf had already declared the Community's support for the ESF.[96] Schuster also confirmed this stance in the meeting of the ESF Preparatory Group at Abingdon in May 1973, by saying that the EC would be prepared to go along with a wider science foundation and abandon its own foundation project restricted to the Community's member states. However, Schuster's remark that it would be impossible for the Commission to state publicly that it would not create a parallel EC organization is intriguing,[97] and it might be explained by the skepticism of some Community member governments toward the foundation as it was now being planned outside the Community: the Belgian representatives in the EC Council had especially been asking whether such a foundation was necessary for assuring European cooperation in fundamental research. They also pointed out that it would be difficult for the EC to create rapports with an institution of nongovernmental nature. Last but not least, a new body like the ESF could eventually supplant the EC institutions—a concern that the Italians also shared.[98]

This unusual situation where the Commission was more inclined toward the form of ESF that now was emerging than some governments of the Community member countries, which continued to express their support for the Commission's vision of the foundation as an EC body, pushed the Brussels institution into an awkward political position: while trying to find footage in the foundation plans, it struggled to retain its credibility as a promoter of a common research policy within the Nine. Pressure only increased when R. St. J. Henry Walker, the secretary of the British Science Research Council and the host of the Abingdon meeting, called for the Community to produce a paper explicitly declaring that the Community welcomed an ESF on the scale proposed by European research councils and academies. The resolution should also affirm the readiness of the EC to write off its plans for a foundation of nine countries in case the project of a larger framework proved successful. From Walker's point of view, the EC's activities in fundamental research could be realized within a larger ESF.[99]

For a Community research policy, the contours of which remained fuzzy and contested, the ESF now appeared a rival that limited the options available for an overall research policy. With the kind of division of labor suggested by Walker, EC activity would have to be restricted to applied and market-oriented research, while the new foundation would be given the monopoly in fundamental research. Moreover, it seemed that the advocates of the non-Community foundation had hijacked the most promising area of activity: experience had already shown how difficult cooperation was in sectors close to the market, where economic interests entered into play.[100] With regard to fundamental research, commercial pressures were less prominent, and chances for successful cooperation were therefore higher. In other words, an ESF outside the EC realm and with an exclusive role in fundamental research would radically alter the prospects for the Community's general research policy and effectively curtail the EC's room for maneuver in the field.

The proponents of a wider ESF were nevertheless ready to welcome Brussels to participate in the setting up of the foundation. This is mainly because the project had yet to be approved at a political level. Of those involved in the preparatory discussions, the Commission was the only institution in the position to speak directly to the national governments that still remained doubtful of the initiative.[101] The Commission was thus useful for acquiring the political support necessary to make the foundation a reality. But there were also some other reasons for keeping the Brussels executive on board: a memo prepared by the British Department of Education and Science stressed the difficulty of drawing a firm boundary between pure and applied research. Although the Department opposed "any attempt by the Commission to assume a coordinating role in regard to basic research," tying the Brussels institution in with the ESF could avoid possible muddles in the future resulting from the entanglement of the different categories of activity. However, this was a practical matter and eventually possible "simply by the Commission having an observer at the ESF."[102]

There was also a third factor supporting the Commission's participation: involving Brussels could be a clever strategy to tie its hands. The Foreign Office toyed with the idea of tying the Commission "fully in with the ESF in order to make certain that it did not seek to operate unilaterally in the field of fundamental research." Rejecting the Commission's financial contribution could lead the Brussels executive to launch some parallel activity.[103] This was the factor that from the British point of view made a financial contribution from the Community acceptable. Yet it should not be too high "since this would involve a risk of the Commission dominating the Foundation proceedings."[104]

While the Commission showed remarkable flexibility with the ESF, it would not be content with any kind of role. One of the Commission's goals was to have a special status that would be clearly different from the position of other participating organizations, such as the OECD or the Council of Europe.[105] This desire to differentiate the EC was basically satisfied at the conference of Gif-sur-Yvette, with the decision to establish a special relationship between the EC and the ESF. According to this agreement, the Commission would make a modest contribution to the administrative costs of the foundation and participate in specific projects, while Commission assessors were welcomed in the ESF Executive Committee meetings.[106] The fears of the Community being relegated to a position similar to the Council of Europe were entirely removed in the second ESF conference in Stockholm in May 1974, where the participants approved an explicit declaration that the Council of Europe would not have equal rights with the EC.[107] Hubert Curien, the French délégué général à la recherche scientifique et technique and the president of the ESF Founding Committee, later explained to the representatives of the Council of Europe that the EC's special status was justified by the unique character of the Community.[108]

The battle over the status within the ESF demonstrates the openness of European integration in the early 1970s. Upset about the EC's creep into its traditional areas of activity, the Council of Europe saw the little Europe of the Six as an increasingly unequal rival.[109] Reacting to a draft resolution, which the Council of Europe circulated just before the EC Paris Summit of October 1972, the French revealingly noted,

> Curiously enough, the Council of Europe is an enthusiastic supporter of the European integration of the Nine within the framework of the Community institutions. There is no doubt that it is because it [the Council of Europe] has the feeling, justifiably, that it cannot compete with it [the European Community], and that it thus has the advantage of placing itself above [the Nine], considering that the construction of the Community evolves within the framework of a wider European cooperation, of which it can be the orienting force or at least the moral leader.[110]

The struggle over the ESF must thus be seen in this context of the fundamental reconsideration of the roles and identities of these two organizations.

Like COST, the ESF came into being as an uneasy compromise that the EC could not but accept. In the Council of Ministers meeting of November 15, 1973, most national delegations welcomed the establishment of relations between the Community and the foundation. However, the representatives also underlined that the foundation should not substitute the Community institutions in the coordination of research policies of the EC member states or in the proposition of Community programs in research.

Moreover, the Community would contribute financially to the foundation only if the latter was already financed by national governments.[111]

Yet there is no denying that a non-Community ESF could be useful for the EC. Through active participation in different European organizations, the EC gained new energy, fresh ideas, actor qualities, and information. Moreover, this activity could boost the Community's visibility, prestige, and even legitimacy. An autonomous ESF led by national research councils and academies was more likely to gain wide popular support than an institution imposed and closely controlled by Brussels. The fact that the ESF could be sold as a grassroot initiative is particularly salient in the context of the reduction in public support for the Community project in the early 1970s, as well as the increasing calls for more social accountability and openness in scientific research.

Importantly, the ESF was also an explicit attempt of the Community to establish itself in humanities and social sciences. The first move into that direction had been the reorganization of the directorial generals and the appointment of Ralf Dahrendorf, a social scientist, to head the new DG Research and Education, whose activity would no longer be restricted to research close to industry and markets. Social sciences and humanities formed an important part of ESF's activities already in its second year of existence.[112] But the establishment of the foundation outside the EC realm strengthened the tradition of organizing European cooperation in humanities and social sciences in forums other than the Community. Researchers in these fields soon recognized the ESF as a desirable framework for promoting cooperation: already before the foundation was even officially set up, British scholars expressed their concerns about what they saw as an insufficient role for human sciences or humanities in the new institution.[113]

The moment when the ESF came into being and the EC Council adopted the Community's action program for research coincided with a change in the general mood within the Community. Now there was a common sentiment that political differences between the member states were growing, not diminishing, and that the willingness of the Nine to overcome the conflicts over the Community's objectives was fading. From this perspective, the years 1969–1972 appear to be a particular momentum, a window of opportunity to build a Community with some important supranational ingredients. It was then the political reality started to

catch up with some of the EC's aspirations of grandeur. The first enlargement round, the creation of the European Council, the gradual extension
of the Community's competences, not to mention the fundamental transformations in transatlantic relations, made it clear that the Community
had established itself as the primary framework of European cooperation.
However disputed the immediate success of the summits of The Hague of
1969 and Paris of 1972 might be, these two meetings clearly demonstrated the willingness of the European decision-makers to give the hypo-
thetic EC "Europe" a more concrete content.[114] The 1970s was a
creational moment, "with (almost) everything in play at the same time,
and a fluid morphing of established categories into as yet undetermined
possibilities, paradigms, and priorities." Federico Romero has highlighted
similarities between this period and the formative era of integration, characterized by constant reassessment of challenges, frenetic comparison and
revision of options, and the complex and often unpredictable ways solutions were reached.[115] This remarkable sense of openness and possibility is
also evident in the research policy debates in the first part of the decade.

At the same time, research was never the priority of the Six/Nine. The
major debates on integration between 1969 and 1972 centered on other
issues, such as economic and monetary policies[116] and political cooperation.[117] Research was not the best topic to attract voters, and many national
politicians felt little enthusiasm for the issue. As a consequence, the success
of the Commission's plans depended on their position in a wider political
context and the support of a handful of key actors of the national science
administrations and research institutions, who nevertheless often preferred
activity outside the Community framework. It is very telling how in 1976,
the Commission was forced to abandon its proposal for organizing a
"European Year of Science," not only because the United Nations already
had similar plans but also because "the scientific community was not very
responsive to the idea."[118]

Despite the common rhetoric of science as international by nature, the
national research systems differed in fundamental ways. After World War
II, though increasingly based on the same vision of economic benefit of
science, research policies in Europe had taken national paths. Adjusting a
number of different practices, institutional structures and traditions would
constitute a major challenge for a common research policy. This was also
something that the creators of the ESF were very well aware of. The first
annual report noted: "Given the existence in western Europe of a wide
variety of institutions for the advancement of research, and of national

science policies, a strongly coordinated supra-national planning authority would be as impossible to realize in this field as it is undesirable."[119] From a political point of view, there seemed to be more diversity than unity in the Western European research landscape.

In the final analysis, competition superseded cooperation. Michael Felder sees the reason for the failure to create a new programmatic basis for European research first and foremost as a result of deep-rooted nationalism that only was strengthened during the years of the economic crisis: "Under the pressure of global economic crisis processes, there was first a clear prioritization for national R&D. The structural policy race among the western industrialized countries took place not only between the three major economic regions, but also between the Western European countries."[120] In this situation, the Nine were not willing to engage in any activity likely to lead to a loss of competitive advantage.

But notwithstanding these challenges, by the late 1970s, the EC had emerged as a new actor in the field of research. The existence and the struggle over the vague objective of a common research policy had significantly boosted the EC entity process, quietly enfolding through successive setbacks and quarrels, as much as through the deceptive moments of consent and unity. Moreover, the Community had contributed to the evolution of a certain set of ideas that the continuous debates had transformed to established premises. Across Western Europe, research was now seen as an important factor behind economic growth and research policy as an activity largely driven by economic concerns. Despite the political revisions of the late 1960s and early 1970s, leading to a greater emphasis on the qualitative aspects of growth and the social accountability of research, growth, accompanied by the strengthening rhetoric of competitiveness, remained the backbone of research policy in Europe. The Community crucially furthered the cementing of this conception of research and research policy that was to form its ideational basis also in the following decades.

NOTES

1. N. Piers Ludlow, *The European Community and the Crises of the 1960s: Negotiating the Gaullist Challenge* (London and New York: Routledge, 2006): 176–177.
2. Ralf Dahrendorf (alias 'Wieland Europa'), "Über Brüssel hinaus," *Die Zeit*, July 9, 1971. Quoted in: Heinrich Siegler, *Europäische politische*

Einigung II (1968–1973) (Bonn, Wien and Zürich: Siegler & Co. Verlag für Zeitarchive GmbH, 1973), 182. The author's translation from German.

3. Ludlow, *The European Community*, 186.

4. For different interpretations on the significance of the meeting see e.g.: Derek W. Urwin, *The Community of Europe: a History of European Integration since 1945* (London and New York: Longman 1995), 157–158; Katrin Rücker, "Willy Brandt, Geoges Pompidou et le sommet de la Haye en 1969," In *Willy Brandt und Frankreich*, eds. Horst Möller and Maurice Vaïsse (München: Oldenburg, 2005), 183.

5. Ludlow, *The European Community*, 192–194; Bino Olivi, *L'Europe difficile. La construction européenne* (Paris: Gallimard, 2007), 105.

6. Ludlow, *The European Community*, 197.

7. Communiqué final du sommet de La Haye (2 décembre 1969). Bulletin des Communautés européennes, Janvier 1970, n° 1. Luxembourg: Office des publications officielles. Quoted in: http://www.cvce.eu/obj/communique_final_du_sommet_de_la_haye_2_decembre_1969-fr-33078789-8030-49c8-b4e0-15d053834507.html (accessed April 19, 2019). The author's translation from French.

8. Ibid.

9. Commission des Communautés européennes: Note de la Commission au Conseil sur les suites à donner au § 9 du communique de La Haye relatif au développement technologique de la Communauté, Strasbourg, le 17 juillet 1970. 19850664, art. 2, ANF; Commission of the European Communities: *Fourth General Report on the Activities of the Communities 1970* (Brussels-Luxemburg, February 1971), 182–183.

10. Commission of the European Communities, *Fourth General Report*, 184–185.

11. Commission des Communautés européennes: Note de la Commission au Conseil concernant une action communautaire d'ensemble en matière de recherche et de développement scientifique et technologique. Supplément 1/71—Annexe au Bulletin des Communautés européennes 1/1971.

12. Ibid.

13. Ibid.

14. Commission des Communautés européennes (Secrétariat du Comité de politique économique à moyen terme): La Communauté européenne et la coopération technologique, Bruxelles, le 8 février 1971. BAC 25/1983-1805.

15. N. Piers Ludlow, "An Opportunity or a Threat? The European Commission and the Hague Council of December 1969." *Journal of European Integration History*, 9 (2003), 11–25.

16. Note à l'attention de mm. des membres de la Commission. 138ème session du Conseil Euratom/Technologie—16/17 décembre 1970, Bruxelles, le 23 décembre 1970. BAC 23/1979-829, CAB.

17. Le Conseil de la CE: Aide-mémoire de la réunion du Groupe de Hauts fonctionnaires de la recherche scientifique tenue les 18 et 19 septembre 1969, Bruxelles, le 22 octobre 1969. 19900634, art. 316, ANF; Note relative à la position du gouvernement Belge au sujet de l'approche globale d'une politique scientifique et technique dans la Communauté, le 31 mars 1971. BAC 23/1979-829, CAB.

18. Communautés européennes, Le Conseil. Note: Coopération européenne dans le domaine de la recherche scientifique et technique (Suite à donner aux rapports des sept Groupes d'experts), Bruxelles, le 20 juillet 1970. BAC 23/1979-825, CAB.

19. Représentation Permanente de la France auprès des Communautés européennes à Monsieur Jean-René Bernard, Conseiller Technique à la Présidence de la République, Bruxelles, le 2 decembre 1970. 19850664, art. 2, ANF.

20. Ibid.

21. Mémorandum sur les modalités d'un renforcement de la coopération européenne en matière de développement industriel et scientifique, le 18 mars 1970. 19850664, art. 2, ANF.

22. Innerer Ausbau der Gemeinschaften: Französisches Memorandum vom 21 März 1970, Bonn, den 3. April 1970. B 35-224, PA AA.

23. Laurent Warlouzet, *Le choix de la CEE par la France. L'Europe économique en débat de Mendès France à de Gaulle (1955–1969)* (Paris: Comité pour l'histoire économique et financière de la France, 2011): 405.

24. John Walsh, "French Science Policy: Problems of Leveling Off." *Science*, March 31, 1972.

25. Hubert Zimmermann, "Western Europe and the American Challenge: Conflict and Cooperation in Technology and Monetary Policy, 1965–1973," *Journal of European Integration History*, 6 (2000): 95–96.

26. Warlouzet, *Le choix de la CEE par la France*, 484–485.

27. De Guazuggli-Marini à Glaesner, le 28 avril 1970. BAC 23/1979-824, CAB. The author's translation from French.

28. Communautés européennes, Le Conseil: Aide-mémoire de la réunion du groupe ad hoc "Recherche scientifique et technique" tenue le 4 décembre 1970 à Bruxelles, Bruxelles, le 4 décembre 1970. BAC 23/1979-828, CAB.

29. A European policy for research and development. Note by J.A. Robinson, July 11, 1972. PRO/FCO 55/908, TNA.

30. Communautés Européennes, Le Conseil: Eléments de l'avant projet de rapport du groupe ad hoc "Recherche scientifique et technique" au

Comité des représentants permanents (texte établi à la suite de la réunion du 19 juillet 1971), Bruxelles, le 26 aout 1971. 19850664, art. 2, ANF.

31. Commission des Communautés européennes, DG III (XII/B): Note: Rôle et structure du CERD. BAC 25/1983-1805, CAB. Note à l'attention de Monsieur Spinelli: Suite à donner au dernier Conseil des Ministres de la Recherche. Esquisse d'un plan par étapes, le 6 janvier 1971. BAC 23/1979-829, CAB.

32. Commission des Communautés européennes, DG III (XII/B): Note: Rôle et structure du CERD. BAC 25/1983-1805, CAB.

33. Préparation du Conseil des 16 et 17 décembre 1970: Conclusions préliminaires du Groupe d'Experts sur la vue en matière de recherche et de développement technologique, Bruxelles, le 11 décembre 1970. BAC 23/1979-828, CAB.

34. Quoted in Luca Guzzetti, *A Brief History of European Union Research Policy* (Luxembourg: European Communities, 1995): 48.

35. Commission des Communautés européennes. Note à la Direction Générale des relations extérieures: Première réunion des experts à haut niveau sur les nouveaux concepts des politiques de la science, Paris, le 2 février 1970. BAC 3/1978, HAEU. The author's translation from French.

36. Reinhard Steurer, *Der Wachstumsdiskurs in Wissenschaft und Politik: Von der Wachstumseuphorie über "Grenzen des Wachstums" zur Nachhaltigkeit* (Berlin: Verlag für Wissenschaft und Forschung, 2002): 422–423.

37. Jan-Henrik Meyer, "Appropriating the Environment. How the European Institutions Received the Novel idea of the Environment and Made it Their Own," *KFG Working Paper* 31 (2011): 24.

38. Steurer, *Der Wachstumsdiskurs,* 424–425.

39. Laura Schichilone, "The Origins of the Common Environmental Policy. The Contributions of Spinelli and Mansholt in the ad hoc Group of the European Commission, 1969–1972," In *The Road to a United Europe. Interpretations of the Process of European Integration,* eds. Morten Rasmussen and Ann-Cristina L. Knudsen (Euroclio, Brussels: P.I.E. Peter Lang, 2009), 336–339.

40. Meyer, "Appropriating the Environment," 5–6, 11.

41. Commission des Communautés européennes: Note de la Commission au Conseil concernant une action communautaire d'ensemble en matière de recherche et de développement scientifique et technologique. Supplément 1/71—Annexe au Bulletin des Communautés européennes 1/1971; Note de Altiero Spinelli aux Membres de la Commission des Communautés européennes, le 5 juillet 1970. AS-00270, HAEU.

42. Laura Grazi and Laura Schichlone, "Environmental Issues in the Improvement of Living and Working Conditions. Innovative Elements in the Process of European Integration during the 1970s," in *Les trajectoires*

de l'innovation technologique et la construction européenne = *Trends in technological innovation and the European construction*, eds. Cristophe Bouneau, David Burigana and Antonio Varsori (Brussels and New York: P.I.E. Peter Lang, 2010), 58, 61–65, 71.

43. Meyer, "Appropriating the Environment," 13–15.

44. Extrait du procès-verbal de la 127ème session du Conseil tenue à Luxemburg le 13 octobre 1970. CM2 1100/b, HAEU.

45. This point was raised for instance by the Commission representative in the ad hoc group "Recherche scientifique et technique" in May 1971. Communautés européennes (Le Conseil): Rapport du groupe ad hoc "Recherche scientifique et technique" au Comité des représentants permanents, Bruxelles, le 26 mai 1971. BAC 23/1979-830, CAB.

46. Extrait, Conseil 123, den 23. Juli 1970. Europäische Zusammenarbeit auf dem Gebiet der wissenschaftlichen und technischen Forschung. BAC 23/1979-825, CAB.

47. Guzzetti, *A Brief History,* 48.

48. Commission des Communautés européennes: Note de la Commission au Conseil concernant une action communautaire d'ensemble en matière de recherche et de développement scientifique et technologique. Supplément 1/71—Annexe au Bulletin des Communautés européennes 1/1971.

49. Politique industrielle, technologique, régionale et environnement au sommet. Note de Christopher Layton, Bruxelles, le 16 novembre 1971. AS-00270, HAEU. The author's translation from French.

50. Ibid. The author's translation from French.

51. Commission of the European Communities: *Sixth General Report on the Activities of the Communities 1972* (Brussels-Luxembourg, February 1973), 219.

52. Forschungs- und Entwicklungspolitik: Vorschlag der EG-Kommission zur schrittweisen Verwirklichung einer gemeinsamen Politik auf dem Gebiet Forschung und Entwicklung der Gemeinschaft. Bezug: Ressortbesprechung in BMBW am 17. November 1972. B 73-431, PA AA.

53. Cited in: Desmond Dinan, *Europe Recast. A History of European Union* (Houndsmills, Basingstoke: Palgrave Macmillan, 2004), 142–143; Urwin, *The Community of Europe,* 158.

54. Dinan, *Europe Recast,* 143.

55. Bruxelles à Ministère des Affaires étrangères, 11.10.1972. 3787 Coopération politique, AMAE. The author's translation from French.

56. Urwin, *The Community of Europe,* 159.

57. Communautés européennes (Le Conseil): Rapport du groupe ad hoc "Recherche scientifique et technique" au Comité des représentants permanents, Bruxelles, le 26 mai 1971. BAC 23/1979-830, CAB; Aufzeichnung für Herrn Spinelli, Mitglied der Kommission: Erste Analyse

der Reaktion auf die Vorschläge der Kommission für eine Aktion Communautaire d'Ensemble (Dokument der Kommission für den Rat vom 11.11.1970, M. Michaelis), den 6. Dezember 1970. BAC 23/1979-828, CAB, Telex de Bruxelles à Paris: Politique commune de la recherche et du développement, 25.7.1972. 19900634, art. 315, ANF.

58. Report from the Working Party on Scientific and Technical Research to the Permanent Representatives Committee, Brussels, April 13, 1973. CM2 1534.1, HAEU.

59. Entwurf einer Weisung für den AStV: Bericht der Gruppe Wissenschaftliche und technische Forschung an den AStV. B 73-7, 431, PA AA; Arbeitsprogramm für den Bereich Forschung, Wissenschaft und Bildung, den 28. Mai 1973. B 73-7, 431, PA AA.

60. Besprechung Bundesminister von Dohnanyi mit Kommissar Spinelli am 10. Juli 1972, Bonn, den 11. Juli 1972. B 35-430, PA AA.

61. Ralf Dahrendorf, interview by Wolf D. Gruner, London, September 2, 1998. Voices of Europe, European Oral History, HAEU.

62. Ralf Dahrendorf (alias 'Wieland Europa'), "Ein neues Ziel für Europa." *Die Zeit* July 16, 1971. Quoted in Siegler, *Europäische politische Einigung*, 185; Ralf Dahrendorf, interview by Wolf D. Gruner den 2 September 1998. Voices of Europe, European Oral History, HAEU.

63. Arthe van Laer, "Vers une politique de recherche commune. Du silence du Traité CEE au titre de l'Acte Unique," in *Les trajectoires de l'innovation technologique et la construction européenne = Trends in technological innovation and the European construction*, eds. Cristophe Bouneau, David Burigana, and Antonio Varsori (Brussels and New York: P.I.E. Peter Lang, 2010): 59. Commission CE: Pour la création d'une assise industrielle européenne. Communication de la Commission relative au programme de politique industrielle et technologique. Supplément n°7 au Bulletin des Communautés européennes 14 (1973); Commission CE, Huitième rapport général sur l'activité des Communautés en 1974, 1975, 297.

64. Van Laer, "Vers une politique de recherche commune," 60; Bulletin des Communautés européennes 14 (1973).

65. Laurence Jourdain, *Recherche scientifique et construction européenne. Enjeux et usages nationaux d'une politique communautaire* (Paris: L'Harmattan, 1995): 55–56.

66. Ralf Dahrendorf, "Recherche, science et éducation," *Information scientifique et technique*, 23.5.1973. The author's translation from French.

67. Débats du Parlement européen. Séance du jeudi 25 novembre 1973, 232.

68. See Conclusion and Further Thoughts.

69. Quoted in Michel André, "L'espace européen de la recherche: histoire d'une idée," *Journal of European Integration History*, 12 (2006): 134.

70. Jourdain, "*Recherche scientifique*," 55–56; Conseil: Programme d'action en matière de politique scientifique et technologique, Bruxelles, le 3 décembre 1973. 19900634, art. 315, ANF.

71. Gemeinsame Forschungspolitik in der EG: Sachstand den 14 Januar 1974. B 73-414, PA AA.

72. Communautés européennes, Comite de la Recherche Scientifique et Technique. Note des Services de la Commission aux Membres du CREST, Bruxelles, le 6 juin 1976. 19850664, art. 3, ANF.

73. There are several and very different interpretations of the origins of the foundation idea. While some maintain that the foundation was first sug- gested by the Dutch and Scandinavian research councils and certain statesmen in other European countries, others stress the role of the OECD in the launching of the initiative. Also, the European Commission officials, such as Christopher Layton, have been identified behind the concept. EFSF Preparatory Group: Resume of the developments leading to the proposal for a European Science Foundation, July 20, 1973. BAC 111/1988-857, CAB; Council of Europe, Consultative Assembly. Committee on Science and Technology. The Project for the creation of a European Science Foundation. Preliminary Draft Report presented by Mr. Boulloche, Rapporteur. Strasbourg, January 7, 1974. BDT 0111/88-865. CAB; Pavitt 1971/1972, 210–273. The Foundation was also elaborated by Pierre Piganiol, the French Délégue général à la Recherche Scientifique et Technique. Pierre Piganiol, "Scientific Policy and the European Community," *Minerva*, 6 (1968), 354–365.

74. EFSF Preparatory Group: Resume of the developments leading to the proposal for a European Science Foundation, July 20, 1973. BAC 111/1988-857, CAB; European Science Foundation: First Annual Report 1974/1975, October 1975. ESF-876, CAB; Hubert Curien, "Vers une Fondation européenne de la science," *Le courrier du CNRS*, 13 (1974), 3–4.

75. Council of Europe. Consultative Assembly. Committee on Science and Technology: Project for the creation of a European Science Foundation. Preliminary draft report presented by Mr. Boulloche, Rapporteur. BDT 0111/88-865, CAB. The interest of the Council of Europe in the Foundation must be viewed against its long-term activity in the field of culture and higher education: since its establishment in 1949, education had been at the heart of the organization's activities. In 1962, with the creation of the European Council of Cultural Cooperation, promotion of European cooperation in science and education was given even more emphasis. Although the Council's cultural and scientific program did not include research activities as such, it nevertheless played an important role in encouraging international exchanges between European universities

and non-university research centers. Joséphine Brunner, "Le Conseil de l'Europe à la recherche d'une politique culturelle européenne 1949–1968," in: *Building a European Public Sphere. From the 1950s to the Present*, eds. Robert Frank et al. (Brussels: P.I.E. Peter Lang, 2010): 164; Jean-Jacques Salomon: "International Scientific Organisations," in *Ministers Talk About Science*, 57, 75.

76. For instance, the OECD envisaged an institution, modeled along the lines of the US National Science Foundation, with a budget representing 10 percent of all national research budgets. Council of Europe, Consultative Assembly. Committee on Science and Technology. The Project for the creation of a European Science Foundation. Preliminary Draft Report presented by Mr. Boulloche, Rapporteur, Strasbourg, January 7, 1974. BDT 0111/88-865, CAB.

77. Brian Flowers, "La Fondation européenne de la science." *Le courrier du CNRS*, 24 (1977), 46–53.

78. Premier Ministre, État-major General de la Défense Nationale, Affaires Économique: Projet O.T.A.N. de création d'un Institut International des Sciences et de la Technologie: Présentation de la proposition Killan, Paris, le 31 octobre 1961. 19920548, art. 20, ANF. Also, Christopher Layton spoke about an executive agency "with substantial budget." Christopher Layton, "Technology and Industry," in *Europe Tomorrow. Sixteen Europeans Look Ahead*, ed. Richard Mayne (London: Fontana/Collins; Chatham House PEP, 1972), 252–253.

79. Council of Europe, Consultative Assembly. Committee on Science and Technology. The Project for the creation of a European Science Foundation. Preliminary Draft Report presented by Mr. Boulloche, Rapporteur, Strasbourg, January 7, 1974. BDT 0111/88-865. CAB.

80. Communication de la Commission au Conseil: Objectifs et moyens pour une politique commune de la recherche scientifique et du développement technologique, Strasbourg, le 14 juillet 1972. CM2 1530.1, HAEU; EFSF Preparatory Group: Resume of the developments leading to the proposal for a European Science Foundation, July 20, 1973. BAC 111/1988-857, CAB.

81. E.G.Willan to Mr. Garvey, June 30, 1967. PRO/FCO 55/43, TNA.

82. Telex de Londres à Paris, le 9 décembre 1972 (No 5332/36). 19900634, art. 315, ANF.

83. Council of Europe, Consultative Assembly. Committee on Science and Technology. The Project for the creation of a European Science Foundation. Preliminary Draft Report presented by Mr. Boulloche, Rapporteur, Strasbourg, January 7, 1974. BDT 0111/88-865, CAB.

84. Austria, France, Germany, Great Britain, Ireland, Italy, the Netherlands, Spain, Sweden and Switzerland.

85. The Commission was represented by Ralf Dahrendorf and Günter Schuster. Co-operation by West European Science Organisations. Report on a meeting held in Munich on April 13–14, 1973 by the British participants. BAC 111/1988-855, CAB.

86. Or European Fundamental Science Foundation as it first was called.

87. Austria, Belgium, Denmark, France, Federal Republic of Germany, Greece, Ireland, the Netherlands, Norway, Portugal, Spain, Sweden, Switzerland, United Kingdom, and Yugoslavia.

88. European Science Foundation. Conference at Gif-sur-Yvette (France): September, 24/25 1973. BDT 0111/88-865. CAB.

89. Suggestions for cooperation in fundamental science between scientific organizations of Western Europe. Drafted by H. Bloch, H.B.G Casimir, R.W.J. Keay and R. St. J. Walker. Enclosure to the letter from Friedrich Schneider, Generalsekretär der Max-Planck-Gesellschaft, Munich, den 26. März 1973. BDT 0111/88-855, CAB.

90. Vermerk für Herrn Professor Dahrendorf: Tagung der 'Preparatory Group' für die ESF in Cumberland Lodge am 20. Juli 1973, Brüssel, den 24. Juli 1973. BAC 111/1988-857, CAB.

91. Günter Schuster, physician, General Director 1973–1981, had his background in Euratom and in the nuclear division of the German Ministry of Research.

92. The author's translation from German.

93. Vermerk für Herrn Kommissar Dahrendorf: Europäische Wissenschaftsstiftung für Grundlagenforschung (EFSF). G. Schuster, Brüssel, den 28. Mai 1973. BAC 111/1988-856, CAB, Suggestions for cooperation in fundamental science between scientific organizations of Western Europe. Drafted by H. Bloch, H.B.G Casimir, R.W.J. Keay and R. St. J. Walker. Enclosure to the letter from Friedrich Schneider, Generalsekretär der Max-Planck-Gesellschaft, Munich, den 26. März 1973. BAC 111/1988-855, CAB.

94. Dahrendorf was, for instance, a personal friend of Dr. Schneider who vehemently opposed Commission participation in the ESF. European Science Foundation, extract from note in informal tripartite talks between United Kingdom, France, and Germany. Smf 17/507/4. PRO/FCO 55/1047, TNA.

95. Note concerning the development of a common policy in the field of fundamental research for the European Communities and the attitude of the European Commission toward the establishment of the proposed European Fundamental Science Foundation, draft May 31, 1973. BAC 111/1988-857, CAB.

96. Co-operation by West European Science Organisations. Report on a meeting held in Munich on 13–14 April 1973 by the British participants. BAC 111/1988-855, CAB.

97. Ibid.

98. Council of Europe. Consultative Assembly. Committee on Science and Technology: Project for the creation of a European Science Foundation. Preliminary draft report presented by Mr. Boulloche, Rapporteur. BDT 0111/88-865. CAB; Conseil: Programme d'action en matière de politique scientifique et technologique, Bruxelles, le 3 décembre 1973. 19900634, art. 315, ANF.

99. Proposed European Fundamental Science Foundation. Notes of Preparatory Group meeting held at Abingdon, 10 and 11 May 1973. BAC 111/1988-856, CAB; Compte-rendu sommaire de la réunion du "Preparatory Group" de la "Proposed European Science Foundation" du 11/12 mai à Abingdon, Berks, Bruxelles, le 14 mai 1973. BAC 111/1988-856, CAB.

100. Michael Felder, *Forschungs- und Technologiepolitik zwischen Internationalisierung und Regionalisierung* (Studien der Forschungsgruppe Europäische Gemeinschaften, Institut für Politikwissenschaft, Philipps-Universität Marburg, 1992): 83.

101. Note à l'attention du Dr. G. Schuster, Directeur Général Recherche, Science et Education: Résumé d'une conversation entre Mr. R. St-J Walker, Secretary of the Science Research Council et M. Romano à Londres le 30 mai 1973, Bruxelles, le 4 juin 1973. BAC 111/1988-857, CAB.

102. Note of Department of Education and Science (P.R. Odgers), October 12, 1972. PRO/FCO 55/1047, TNA.

103. Ibid.

104. European Science Foundation, extract from note in informal tripartite talks between United Kingdom, France, and Germany. Smf 17/507/4. PRO/FCO 55/1047, TNA.

105. To Mr. Schuster. BAC 111/1988-865, CAB; Suggestions for CCE positions on redrafting the status of the ESF, Note of Michael Proctor, May 4, 1974. BAC 0111/1988-865, CAB.

106. CNRS: European Science Foundation: Conference of representatives of European science councils and academies held at the Centre National de la Recherche Scientifique, Gif-sur-Yvette, France, 24–25 September 1973. BAC 111/1988-859, CAB.

107. European Science Foundation—Conference at Stockholm May 2 and 3 1974, Note of Michael Proctor, May 4, 1974. BAC 111/1988-865, CAB.

108. Procès-verbal, de la réunion du Founding Committee de l'ESF sous la présidence de Monsieur H. Curien, CNRS, Paris, le 6 juillet 1974. ESF-864, 001:06(4), CAB.

109. Brunner "Le Conseil de l'Europe," 166.

110. Sous-Direction d'Europe Occidentale. Note pour le cabinet du ministre délégué, le 14 octobre 1972. 3787 Coopération politique, AMAE. The author's translation from French.

111. Communautés européennes, le Conseil: Note: Programme d'action en matière de politique scientifique et technologique, Bruxelles, le 21 décembre 1973. BDT 0111/88-865. CAB.

112. Flowers, "La Fondation européenne de la science," 46–53.

113. *The Times,* September 27, 1973; *The Times Higher Educational Supplement,* November 16, 1973.

114. Bo Stråth and Hagen Schulz-Forberg go even further: in their view, the early 1970s marked the point of the culmination of the federalist project. The plans for a European economic and monetary union and a European political cooperation (the Werner and Davignon Plans) contained greater potential for a democratic and political Europe than the subsequent projects revolving around the vague idea of European identity. Bo Stråth and Hagen Schulz-Forberg, *The Political History of European Integration* (Oxon and New York: Routledge, 2010), 2, 12, 14, 36.

115. Federico Romero, "The International History of European Integration in the Long 1970s. A Round-table Discussion on Research Issues, Methodologies, and Directions. Research Report," *Journal of European Integration History,* 17 (2011), 335.

116. For the history of the EMU, see e.g.: Emmanuel Mourlon-Druol, *A Europe Made of Money: the Emergence of the European Monetary System* (Ithaca, N.Y.: Cornell University Press, 2012); Harold James, *Making the European Monetary Union* (Cambridge, Mass.: Harvard University Press, 2012).

117. For European political cooperation, see e.g.: Angela Romano, *From Détente in Europe to European Détente: How the West Shaped the Helsinki CSCE* (Brussels and New York: P.I.E. Peter Lang, 2009).

118. New orientations of EEC R&D policy. Address to the General Assembly of the European Science Foundation (Günter Schuster), Strasbourg October 26–27, 1976. ESF-877, CAB.

119. European Science Foundation: First Annual Report 1974/1975, October 1975. ESF-876, CAB.

120. Felder, *Forschungs- und Technologiepolitik,* 84. The author's translation from German.

PART III

CHAPTER 8

The Return of the Gap

By the early 1980s some of the key economic, political, and technological trends that had been visible along the late 1960s and early 1970s, and that had shaped the early EC research policy, had become palpable: stagnating economic growth, the breakthrough of information and communications technology and other advanced technologies, and accelerating international economic competition in the increasingly open and crowded global markets provided the decisive context for the 1980s' discussions on research. If the research policy debates in the beginning of the previous decade had included some questioning and reassessment as well as openness for broader socio-environmental concerns, now there was a drastic return for a narrow ideational framing, focused on national industries' innovative and productive performance.[1] And the change had longstanding consequences. "This shift in perceptions about the pressing policy needs to redress the structural flaws of Western economies," Egil Kallerud writes, "marks the beginning of modern innovation policy as a framework within which we have since been thinking about science, technology and innovation for economic growth and competitiveness."[2]

The OECD captured these moods with a report *Technical Change and Economic Policy, Science and Technology in the New Economic and Social Context,* prepared by a group of leading experts from industry and universities.[3] A result of two years' work, *Technical Change and Economic Policy* was the first major OECD report on science policy after the influential Brooks Report, *Science, Growth and Society,* which had come out almost a

© The Author(s) 2020
V. Mitzner, *European Union Research Policy,* Europe in Transition:
The NYU European Studies Series,
https://doi.org/10.1007/978-3-030-41395-8_8

decade earlier. It was considered as "the most comprehensive and concise international effort to re-evaluate the links between research, technology, and the economy that can be found today."[4] The report identified three fundamental changes in the world that had created the need to reassess the premises of science policy: the emergence of the problem of world energy supplies because of the OPEC (the Organization of the Petroleum Exporting Countries) crisis, the changed international position of the developing countries, and the slowing down of global economic growth with its concomitant threat to employment.[5]

That growth emerged as the key concern of the new OECD report on research is not surprising. Matthias Schmelzer has shown that toward the late 1970s, the growth paradigm, which had come under attack in the late 1960s and early 1970s, gradually crept back to the center of the organization's agenda and, after the OECD ministerial meeting of 1975, was reaffirmed as the fundamental political goal of industrialized countries.[6] This trend was strengthened by the *Interfutures* report that the organization published in 1979. Developed by Club of Rome founder and OECD director Alexander King and chaired by Japanese economist Saubro Okita, who also was a Club of Rome member, *Interfutures* intended to authoritatively counter the growth critics. The study denied that there would be physical limits to growth and argued that social, economic, and political limits could be overcome by political means, such as a mix of market-oriented structural reforms and environmental regulations. According to Schmelzer, the key conclusion of the report, emphasizing the complementarity and compatibility of environmental and growth-oriented economic policies, soon became the official OECD position.[7]

These discussions within the OECD reflected wider international trends. The victory of Ronald Regan in the 1980 US presidential elections affirmed the status of growth as one of the nation's main political goals. Robert Collins has shown how the 1970s' battle between the three economic currents for the position as the guiding principle of US politics, namely, the traditional demand-side fiscal Keynesianism, an approach supporting micro-economic intervention and guidance through industrial policy, and a new growth-oriented economic analysis and policy focusing on the preeminent contribution of the supply side to economic well-being, resulted in the crushing victory of the latter. According to Collins, the growth-harnessing supply-side approach, which was advocated by conservative politicians and thinkers and increasingly supported by preeminent voices from the academic world, benefited from the inadequacy of the

Keynesian economic theory to explain the economic hurdles of the 1970s, as well as the confusion and controversy around industrial policy. In the conservative circles, economic growth was portrayed as a prize that would follow tax cuts, spending controls, and deregulation. The objective of growth was fully embraced by Reagan, who in 1980 proclaimed that "[t]he Republican program for solving economic problems is based on growth and productivity."[8] Reagan's drive for growth went hand-in-hand with disregard to environmental problems. The president dismantled environmental regulations, cut the budgets of government regulatory agencies, and appointed notoriously anti-environmentalist figures to key environmental positions.[9]

Similar winds were blowing in Europe as well, where national economies were gripped by recession. Between 1974 and 1980, the economic growth in the European Community area declined from 4.6 percent in 1960–1970 to 2 percent, while unemployment increased from two million in 1970 to eight million in 1981. The inflation rates were soaring, exceeding 10 percent on average during 1975–1980 (the average was 3.5 percent in the 1960s), and the burden of public expenditure reached almost 50 percent of the national product.[10] While important political differences persisted between the European countries, curbing public spending and tightening monetary policy to control inflation were widely approved as responses to the sterner economic conditions.[11] The most extreme case was Margaret Thatcher's Britain, where the embrace of austerity was abrupt and radical. Reversing the country's economic decline constituted the core of Thatcher's political design and neoliberal crusade, which involved deep tax cuts, monetary control, privatization of much of the public sector, deregulation of financial market, and a general shrinking of direct government involvement in the economy.[12] In Germany, the arrival in power of Helmut Kohl in 1982 marked a shift of normative emphasis within an already emerging new economic orthodoxy around monetarism and supply-sidism.[13]

Given the conservative turn in the United States and in many parts of Europe, and the following embrace of economic growth as a central goal of national politics, it is not surprising if the research-growth link retained much of its earlier vigor. True to the organization's traditional approach, the OECD expert group that authored the 1980 report on research, shared the assumption that research would be of fundamental importance to national economic performance. "A central task of science and technology policy is to monitor and mold these relations [between technical change and economic progress] so as to reinforce to a maximum the contribution of science and technology to economic and social progress," it concluded. Indeed, the report considered the idea that research and development activities were the motive forces of growth as a "general conviction."[14]

Governments across the Western world were concerned about the dangers that research and innovation faced because of the slowdown of economic growth.[15] Reflecting this changed economic reality, during the 1970s, the growth rate of industry-financed R&D decreased, and industrial R&D became concentrated on short-term low-risk projects. In addition, there was a slowdown and in some cases even an end in the growth of academic research. After a rapid increase of university research expenditure during the 1960s, largely due to increasing university enrolments and public policies boosting university research through research councils and other bodies, after 1971 or 1972 came a period of stagnation in several countries. Although in 1980, the OECD was unable to make any generalizations about the effects of reduced growth on the development of different research fields, the organization called for deliberate policies "to ensure that fundamental science and technology will not be neglected, or the long-term possibilities for economic growth and change impaired by such neglect."[16]

In the early 1980s, the future of research did appear gloomy in several Western European countries. "Things have changed so much in the support of science in this country and others, that the means for creating scientific policy and implementing it are not so readily forthcoming as they were, say 5 or 10 years ago," a British observer declared in an unusual symposium that in 1981 brought together European policy-makers and influencers to discuss science policy "in the new socio-economic context."[17] In the United Kingdom, Prime Minister Thatcher was a firm believer in basic science, but "during the early 1980s," her Science Advisor George Guise recalls, "there was a general Whitehall attitude that too many crazy people were building radio telescopes or atom smashers and that a Thatcherite administration should focus research on utilitarian goals."[18] The subsequent budget cuts led to a widespread discontent within the academic world, and in January 1986, more than 100 fellows of the Royal Society and almost 1500 scientists sponsored a half-page advertisement in *The Times* to launch a movement to save British science.[19] In France, the situation was not much different. Between 1968 and 1981, research spending had fallen from 2.2 percent of the GDP to 1.8

percent.[20] Even in Germany, the research budget was tight. Interviewed for *Bild von Wissenschaft* in October 1981, Andreas von Bülow, the German research minister, complained about the scarcity of the funds he had available, concluding bitterly, "I am barely able to pay all the bills."[21]

These circumstances provided a fertile ground for the renewed fears of a technology gap. During the 1970s, the changes in the discourse of science and technology had served to calm European panic about the gap. Criticism of the generous public R&D spending that in the 1960s had mostly favored large demonstration projects, and the doubts about the credibility of "linear model" of innovation, reduced the political leverage of anxious comparisons between Europe and the United States. At the same time, the economic recession that hit America and the 1967–1977 stagnation in the volume of US government-sponsored R&D made it seem as if the gap was losing its relevance.[22] Tellingly, the 1980 OECD research policy report remained silent about the gap and noted the relative weakening of the American position: whereas in 1967, the United States had performed about 60 percent of the OECD area's industrial R&D, in 1975 its share had gone down to 50 percent. This went together with the decline of the US basic research in industry, which between 1967 and 1975 fell about 30 percent.[23]

However, as the decade went on, the fears of the gap crept back. And this time they appeared more pervasive and threatening than ever. The major novelty was that now the challenge came from two directions: for some time, Japan had been making disquieting leaps in consumer technology and occupying new markets at a stunning speed. At the same time, competition among the Western countries was tightening: in the United States, President Ronald Reagan sought to revitalize the US economy by moving back to the fast-growth track of the earlier decades, and essentially, this growth was harnessed through research. The Reagan years witnessed a substantial increase in the federal support for research. Between 1980 and 1988, defense R&D increased from $15 billion to $40 billion, while the corresponding figures in nondefense basic research funding were $4.2 billion and $8.5 billion. This drive of public politics was strengthened by a simultaneous increase in R&D spending in the private sector[24] that Frank Press, writing in *Science*, called as "something of an R&D renaissance."[25] While a big chunk of this research spending went to the Strategic Defense Initiative (SDI) or Star Wars, a major military program with a projected budget of $26 billion, in Europe, a common perception was that these vast sums of money would enable American contractors to achieve scientific breakthroughs with commercial advantages. The US defense R&D programs in the 1980s reminded Europeans of the earlier National Aeronautics and Space Administration (NASA) and Air Force programs that had made possible the early American advance in crucial technologies, such as integrated circuits (IC) and computers.[26]

Indeed, the electronics industry emerged as a particular cause of concern. In 1980, the OECD considered electronics "by far the ripest and most pervasive technology of the late 1970s."[27] Its potential was significant, because innovations in the field "are bound to have pervasive effects in the many sectors where improved methods of calculation, communication, control, and the storage and manipulation of information are necessary or possible. The diffusion of electronics throughout other manufacturing and service industries will result in an economy in which one technology influences innovation almost everywhere."[28] In other words, the electronics industry held out the promise of growth.

At the heart of the 1980s discussion of electronics was the "telematics" revolution. The word telematics comes from the French neologism *télématique*, coined to combine telecommunications and *informatique* or data-processing. These two originally separate technologies as telephones and adding machines become entangled as a result of the fast development of ICs. Two American firms invented the IC in the late 1950s, and this technology, which contains on a tiny single chip of silicon hundreds of thousands of miniaturized electronic components, soon led to major leaps in several microelectronics sectors, such as televisions, stereos, computers, and telecommunications. "The combination of steady improving performance and constantly declining prices," Wayne Sandholtz wrote in 1992, "has meant that ICs have become increasingly attractive for virtually innumerable applications."[29] By the early 1980s, the IC markets were growing in Europe faster than gross national products (GNPs).[30]

Take the example of computers. The increased speed, capacity, and reliability, as well as the falling prices of ICs, made computers increasingly powerful and inexpensive.[31] The capabilities of one of the first electronic computers (ENIAC), built in the 1940s for several million dollars, could, in the late 1970s, be produced for less than 100 dollars. And this computer calculated twenty times faster, was 10,000 time more reliable, and required 56,000 times less power and 300,000 times less space.[32] Consequently, the number of computer systems installed in the United States climbed from 5500 in 1960, to 400,000 in 1983. By 1988, already 18 percent of American households had a personal computer. Moreover, computer markets in the advanced industrialized countries more than tripled from 1975 to 1982. Similar growth rates were observed in telecommunications equipment, where the world market more than doubled in value from 1977 to mid-1980s.[33]

By 1980, Japanese and American firms dominated both the European and world markets for important high-tech sectors, such as semiconductors and computers. Between 1978 and 1983, the share of West European firms in world semiconductor markets fell from 16 percent to about 12 percent, whereas the share of Western Europe in world production of ICs declined from 6.7 percent in 1978 to 5.8 percent in 1980. The plunge was dramatic. Between 1981 and 1982—within just two years—France's trade deficit in electronics doubled to twelve billion francs. In 1983, nine of the top fifteen computer makers in Europe were American, and the data revenues of the American giant IBM alone were larger than those of the next nine largest manufacturers combined. It truly looked like the fruits of the microelectronic revolution were not benefiting Europeans, as they were their rivals in the East and West: the share of German industry in the domestic market for ICs was only 60 percent in 1984.[34]

Japan's rise had been steady and systematic. Still in the 1950s, the country's main export industries consisted of labor-intensive produces, where low wages enhanced its competitive position vis-à-vis Western industrialized countries. In the early 1960s, Japan moved to engineering industries, and at the end of the decade, Japanese radios, television sets, motor vehicles, and petrochemical products already raked with world's leaders. While Japan's rapid industrial progress was largely based on the acquisition and improvement of existing technological know-how, it soon started to invest in R&D on its own.[35] Domestic consumer industry provided the initial demand for Japanese semiconductor companies, but it did not take long until computer and industrial applications became the main sources of demand for advanced ICs. This led to a successful adventure to foreign markets. Between 1976 and 1981 Japan's trade balance in ICs converted from a $142 million deficit to a $239 million surplus. By 1985 a Japanese company, Nippon Electric Company, had become the world's largest producer of semiconductors, and a year later, Japan's total world market share for IC's had grown to 45 percent, surpassing the United States.[36] "Japanese dynamism is such that she may well become the pace setter, closely followed by the US and Germany,"[37] Daniel T. Jones, a British observer, noted in 1980.

The fears of the gaps were exacerbated by a more competitive economic environment, resulting from the lowering of tariff barriers and increasing interdependence amongst the OECD countries, as well as the emergence of newly industrialized countries as serious players in the world markets. Competing with the OECD countries on the basis of cost over a wider range of standard—and in some cases even sophisticated—manufacturing products and materials, the newly industrialized countries not only pushed the OECD toward more capital-intensive technologies and toward products and materials of higher quality but also created a demand for services targeted at their growing markets.[38] "A strong capacity for innovation," the 1980 OECD report concluded, "will be even more important for competitiveness in manufacturing industry in the future than it was in the past."[39]

Moreover, the increasing cost and scale of research and the continued rise of large multinational enterprises served to limit possibilities for governmental activity. "While much research is still undertaken by universities, cooperative research institutions and small uninational firms," John H. Dunning noted in 1993, "the commercialization of new generic technologies is increasingly requiring a network of physical facilities and organizational and management capabilities which global companies are best equipped to supply. ...For some time now, large industrialized firms have been the main producers, organizers and users of technological capacity." Dunning estimated that in the late 1980s, multinational companies accounted for between 75 percent and 80 percent of the privately undertaken R&D in the world.[40] The research landscape in Western industrialized countries had become increasingly complex and competitive.

The early 1980s, and in particular the 1980 OECD report *Technical Change in Economic Policy*, prepared the ground for a shift from research and science polity to innovation policy. While the term "innovation" had already featured in policy discussions during the earlier decades, toward the late 1980s, it emerged as a central concept framing debates and activities in research. Egil Kallerud describes this change as an "emergence of a new conception of the role of economic objectives in science or R&D policy, following from the acknowledgement of the *fragility* of the dynamics that underpin productivity growth through a high, sustained rate of technical change." This new policy framework "successfully incorporate[d] an increasingly large number of forms and determinants of innovation policy." While consolidating economic growth and competitiveness as core political concerns, the innovation policy framework adopted a wider structural lens, emphasizing the distributive and absorptive capacity of "national systems of innovation." Kallerud sees the 1980 OECD report as articulation of a paradigm shift, "in explicit opposition to the 'social priorities paradigm' of the Brooks report." Indeed, the report included several elements which by the following decade became dominant, such as "the primacy of economic imperatives and objectives, the role of science-based technologies as privileged sources of technological change and innovation, the need to integrate more closely science, technology and economic policies, stronger coordination and alignment between innovation/economic and other policies and policy areas."[41]

Johan Schot and W. Edward Steinmueller also recognize an ideational shift in science policy centered on innovation. According to these two authors, the policy change was motivated by the intensification of international economic and technological competition, the rise of East Asian economies, and the discovery that important national differences might exist in capacity to innovate. In response, scholars started to examine different configurations of organizations involved in generation and use of scientific and technological knowledge. National systems of innovation brought a shift from a linear understanding of innovation toward a more interactive model, emphasizing the intertwined nature of government industry and university research efforts. Instead of flowing linearly from science to applied R&D to commercialization, knowledge was seen to be generated through interaction among various actors, working collaboratively across sectors.[42]

It needs to be said that this continuing obsession of growth and competitiveness did not mean a return to the certainty of the 1950s and 1960s. Writing in *The OECD Observer* in 1980, Jean-Jacques Salomon noticed that the relationship between scientific research, technological development, and economic growth "is not easy to describe because the interactions are complex and not amenable to quantitative measurement with presently available tools." In fact, the only available tool was statistics on investment in research and development, and these statistics only offered a partial picture, since they measured input rather than results.[43] *Technical Change and Economic Policy*, on the other hand, acknowledged that "our ability to assess the effectiveness of various mixed policies on the promotion, development and application of science and technology is still limited. This means that government policies for R&D or, more broadly, for science and technology, are bound to be subjects of controversy."[44] Furthermore, the OECD appeared cautious even about the benefits of growth itself. While fully embracing the goal of economic growth, the organization recognized that the GDP "is not adequate as a general measure of welfare." Indeed GDP "does not even purport to measure certain important kinds of benefits and costs of economic change," such as the lower infant mortality rates, greater freedom of certain kinds of diseases, longer life span, and the enormous improvements in communication and travel. Also, the GDP statistics failed to reflect "many of the social costs and disadvantages associated with economic growth and technical advance," such as pollution, noise, congestion, and unemployment in certain sectors.[45]

Furthermore, the 1980s was also the decade with new consciousness of the dangers of climate change. Whereas in 1981 only 38 percent of Americans had heard of the greenhouse effect, by 1989, the number had leaped to 79 percent.[46] Eventually, this new threat would transform the way people saw their planet and its future existence.[47] But in the 1980s, all this still seemed distant, and the worrying results of science showing that the planet was warming, had little direct impact on economic and research policy. The latter was increasingly steered toward fostering innovation in response to competition in the entangled global economic system. And at the heart of that system was almost universal aspiration for economic growth, which continued to define the political goals in liberal democracies as well as in the trembling socialist regimes. If there was a paradigm change or shift in policy framing in research policy, it remained partial. Finally, with the victory of capitalism in Eastern Europe, the 1980s went into history, not as a decade of embracing the ideas of ecological and social sustainability but as a decade celebrating the Western values of commerce and consumption. In that increasingly fast and interconnected world, innovation policies were hailed as vital, enabling mechanisms, improving national economic performance, and spurring productivity and growth.

NOTES

1. Johan Schot and W. Edward Steinmueller, "Three Frames for Innovation Policy: R&D, Systems of Innovation and Transformative Change," *Research Policy* 47 (2018): 1558.
2. Egil Kallerud, "Goal Conflicts and Goal Alignment in Science, Technology and Innovation Policy Discourse," Working Paper (Oslo: Nordic Institute for Studies in Innovation, Research and Education, 2010) 9.

3. The members were: Bernard Delapalme, A. Caracciolo di Forino, Umberto Colombo, Christopher Freeman, Herbert Fusfeld, Robert Gilpin, Claude Gruson, Albert O. Hirschman, Helmar Krupp, Gösta Lagermalm, Richard Nelson, Keith Pavitt, Gert W. Rathenau, Nathan Rosenberg, and Emma Rotschild; OECD, *Technical Change and Economic Policy. Science and Technology in the New Economic and Social Context* (Paris: OECD, 1980), 5; "Science policy in the new socio-economic context," *Science and Public Policy* (April 1981): 86–91.

4. Jean-Jacques Salomon, "Technical Change and Economic Policy," *The OECD Observer* (May 1, 1980): 16–22.

5. OECD: *Technical Change and Economic Policy*, 13.

6. See also Matthias Schmeltzer, *The Hegemony of Growth. The OECD and the Making of the Economic Growth Paradigm* (Cambridge: Cambridge University Press, 2016): 315.

7. Ibid., 317–20; OECD, *Interfutures. Facing the Future, Mastering the Probable and Managing the Unpredictable* (Paris: OECD, 1979).

8. Robert M. Collins, *More. The Politics of Economic Growth in Postwar America* (Oxford: Oxford University Press, 2000): 167–197.

9. Benjamin Kline, *First Along the River. A Brief History of the U.S. Environmental Movement* (Lanham: Rowman & Littlefield, 2007): 101.

10. Charles S. Maier, "Consigning the Twentieth Century to History: Alternative Narratives for the Modern Era," *The American Historical Review*, 105 (2000), 807–831.

11. David P. Calleo, "Introduction: National Strategies and the 'New Europe'," in *Recasting Europe's Economies, National Strategies in the 1980s*, eds. David P. Calleo and Claudia Morgenstern (Lanham, New York and London: University Press of America, 1990), 8–9.

12. Robert Skidelsky, "Britain, Mrs. Thatcher's Revolution," in *Recasting Europe's Economies, National Strategies in the 1980s*, eds. David P. Calleo and Claudia Morgenstern (Lanham/London: University Press of America, 1990): 101–128.

13. Jeremy Leaman, *The Political Economy of Germany Under Chancellors Kohl and Schröder. Decline of the German Model?* (New York and Oxford: Berghahn Books, 2009), 12–13.

14. OECD, *Technical Change and Economic Policy*, 11–12.

15. Salomon, "Technical Change and Economic Policy," 22.

16. OECD, *Technical Change and Economic Policy*, 43, 94.

17. Michael Brett-Crowther, "Extrapolating and Escaping," *Science and Public Policy*, April 1981, 82–84.

18. George Guise, "Margaret Thatcher's Influence on British Science," *Notes and Records, Royal Society of London*, 68(3) (2014): 301–309.

19. Denis Noble, "We Are Still Saving British Science from Margaret Thatcher," *Nature*, April 17, 2013.

20. *New Scientist*, July 9, 1981.

21. Gespräch mit Forschungsminister Andreas von Bülow. *Bild der Wissenschaft*, October 8, 1981.

22. Bruce L. R. Smith, *American Science Policy since World War II* (Washington D.C.: The Brookings Institution, 1990): 102–104.

23. OECD, *Technical Change and Economic Policy*, 33–37. Similar trend, however, was observed in Japan as well, where the share of R&D expenditure of business enterprises to basic research declined even more drastically: from 10.2 percent in 1967 to 5.2 percent in 1975.

24. Smith, *American Science Policy*, 132–133, 142.

25. Frank Press, "Rethinking Science Policy," *Science*, October 1, 1982. However, in the early 1980s, the United States too was struggling with the economic difficulties, and the Reagan Administration science policy was by no means a blessing for all research sectors. Convinced that the federal government should concentrate on basic research and military R&D and leave commercial technological development to the private sector, the Administration pulled back from several development projects and cancelled innovation programs that had been started by its predecessor. Among the affected activities were research programs in renewable energy and conservation. Also, funding for social sciences and education diminished. Colin Norman, "Reagan's Science Policy," *Science*, November 12, 1982.

26. Wayne Sandholtz, *High-Tech Europe: the Politics of International Cooperation* (Berkley and Los Angeles: University of California Press, 1992): 126.

27. OECD, *Technical Change and Economic Policy*, 55.

28. Ibid., 48.

29. Sandholtz, *High-Tech Europe*, 47.

30. Ibid., 45–49.

31. Ibid., 50.

32. Salomon, "Technical Change and Economic Policy," 16–22 (18).

33. Sandholtz, *High-Tech Europe*, 50–55.

34. Ibid., 44, 71, 124, 146.

35. Roy Rothwell, "Innovation Policy; Reindustrialization and Technology: Towards a National Policy Framework," *Science and Public Policy*, June 1985, 117.

36. Sandholtz, *High-Tech Europe*, 122.

37. Daniel T. Jones, "British Industrial Regeneration: The European Dimension," in *Britain in Europe*, ed. William Wallace (London: Royal Institute of International Affairs, 1980), 122.
38. OECD, *Technical Change and Economic Policy*, 54.
39. Ibid., 95.
40. John H. Dunning, *Multinational Enterprises and the Global Economy* (Wokingham, England; Reading, Mass.: Addison-Wesley, 1993), 290–291.
41. Kallerud, "Goal Conflicts," 10–14.
42. Schot and Steinmueller, "Three Frames," 1558–1561.
43. Salomon, "Technical Change and Economic Policy," 16.
44. OECD, *Technical Change and Economic Policy*, 38.
45. Ibid., 61.
46. Naomi Klein, *This Changes Everything* (Penguin; Random House, 2015), 73–74.
47. The first scientific discoveries showing that burning carbon could be warming the planet stem from the 1950s. However, only in the late 1980s, the climate change became part of a wide public discussion. Klein, *This Changes Everything*, 73. See also Joshua P. Howe, *Behind the Curve. Science and the Politics of Global Warming* (Seattle: University of Washington Press, 2014).

CHAPTER 9

Research Policy: A Trailblazer for Institutional Change

The previous chapters have shown how after the mid-1960s the European Community (EC) gradually established itself as a salient player in the field of research. By the end of the 1970s, it had a specific directorate general for research and a number of research programs administered from Brussels. Moreover, the Commission was deeply involved in several other efforts of European research cooperation, such as the European Science Foundation, whose establishment not only highlighted the continuing tensions surrounding the EC's role in research policy but also underlined the fact that the Community had become an actor to be reckoned with.

It is true that the scope of the Community's activity in research still was relatively limited: the EC budget for 1980 authorized staff of less than 3000 persons for all Community's R&D activities, which was only 0.3 percent of people engaged in scientific and technical work in the EC area. The research budget itself was about 1.5–1.8 percent of total public sector approbations to research in the member states and strongly dominated by the energy sector (about 70 percent of all funds went to energy). However, between 1975 and 1980, the Community's spending on research more than doubled, growing from 70 million ECU to 300 million ECU. In international cooperation, member states increasingly tended to give priority to activity within the Community: between 1974 and 1978, the ratio between Community cooperation and international cooperation had risen from 8.5 percent to 17.3 percent. This was still over five times less than the amounts spent for the European Space Agency (ESA), European

© The Author(s) 2020
V. Mitzner, *European Union Research Policy*, Europe in Transition:
The NYU European Studies Series,
https://doi.org/10.1007/978-3-030-41395-8_9

Organization for Nuclear Research (CERN), International Atomic Energy Agency (IEA), European Molecular Biology Organization (EMBO), and military cooperation, but the trend was clear.[1] The Brussels institution also estimated that "most work conducted has led to significant results." For example, research on new and renewable energy sources had served as an important catalyst for national efforts and encouraged cooperation between industrial companies and laboratories.[2] Last but not least, there seemed to be strong public backing for the Community's research projects: in a poll from 1977, a total of 79 percent of the respondents supported the idea of common activity in the field.[3]

The Commission's assessments paint a rather positive picture of those years. A document from April 1980, looking back to the four previous years, highlighted the gradually increasing interest in this policy area:

> The period 1976–80 proved to be crucial for the definition and implementation of the common R&D policy. After the Euratom crisis and the hesitations of the Member States towards Community research, which characterized the first years of the decade, a new dynamic has developed since 1976. Objectives, clear priorities, and selection criteria have been set, allowing the adoption of several programs of increasing scale and impact. The concentration of efforts on a few key areas has resulted in much closer coordination of national activities. Although this period has not given the common R&D policy the dimension that it eventually should have, it has granted it most appreciable movement and vitality.[4]

Another document, drafted in the following year, pictured the progress in an equally favorable light:

> With the year ending, comes to an end a first phase of six years of implementation of a real common R&D policy; Objectives have been set, decision-making procedures and implementing rules have been defined and made operational, programs have been launched.[5]

With the growing appetite for more ambitious action, the EC's accumulated experience and expertise became a valuable political asset. "It is the experience which past achievements have brought to the Commission," a Commission document from the early 1980s asserted, "which gives it both the right and the justification to suggest a new stage in the progress of European R&D."[6] By the turn of the decade, the EC was firmly on its way to become a powerful arena for European research.

Yet it was only in the early 1980s that research became a full-fledged EC policy with sizeable budgets, truly ambitious programs, and a more strategic approach. So far, the Community's activities had been sectorial and selective, lacking an overall view of the mushrooming initiatives. While relatively steady, this progress was slow and seen as clearly insufficient to match with the rapidly changing political and economic circumstances in Europe. Only the introduction of the major information technology programs and the agreement on the first framework program for research in 1983 marked both a qualitative and a quantitative leap in the Community's activity. These programs not only introduced an entirely new concept for planning and managing the EC's activities, they also led to a substantial budget increase, setting the trend for the subsequent decades.

This last chapter of the book explores the creation of the EC's new political agenda for research, which ultimately proved to be more successful than any of the earlier proposals in the field. Toward the late 1970s, the Commission grew frustrated with the slow progress with its initiatives and changed its strategy: the launch of the first framework program in 1984 as well as the first major Community technology programs (such as ESPRIT) was made in close consultation with researchers and research administrators. Crucially, the Brussels institution could now rely on a sturdy institutional setting. By the 1980s, an "epistemic community" of national and independent actors and the Commission officials had emerged through regular meetings in various Community committees and working groups, the most important of which was CREST (Comité de la recherche scientifique et technique), established in 1974. Moreover, the new overtures drew on earlier initiatives and discussions on research policy. As much as an answer to the transforming economic and technological landscape, they were an extension of an existing policy relying on a consolidated set of beliefs. Technological change and intensifying worldwide competition in R&D would not have created sufficient pressure without a firm conviction that research contributes to economic growth and increasing competitiveness, as well as the resurgence of old fears of Europe's retard vis-à-vis its main competitors. Without this institutional and ideational continuity, the new research policy initiatives would hardly have been so successful.

For anyone familiar with the traditional narratives of EU history, this expansion of the EC research activities might seem surprising: several accounts picture the Community in the early 1980s as being in the middle of a severe economic and political crisis. Many contemporary sources confirm this impression. In 1982, *The Economist* published its iconic cover with a picture of a tombstone and the sinister words: "EEC born March

25th, 1957, moribund March 25th, 1982, capax imperiii nisi imperassett."
(It seemed capable of power until it tried to wield it.)[7] Two years later,
Gaston Thorn, the disheartened Commission president, delivered his
gloomy, often cited, speech "European Union or Decline: To Be or Not to
Be," deploring what he saw as a paralysis of the Community. "The
European Council's repeated failures, the delays, the bickering, the deep-
seated crisis of the Community system, prompt me to ask whether we still
will to integrate, whether that will is strong enough," Thorn questioned.[8]
It really seemed as if the Community had lost its momentum.[9]

The most immediate reason for such pessimism was the prolonged dispute over the British contribution to the Community's budget, which
effectively poisoned the political atmosphere and prevented progress in
almost every other policy area. But there were also other factors, such as
the difficult economic situation in the member countries, strained enlargement negotiations, disagreement over institutional reform, the increase of
the Community's own resources, tensions arising from rampantly increasing spending on agriculture, weakening of the Franco-German concord,[10]
and last but not least, general lack of enthusiasm to proceed with integration.[11] All this changed by the mid-1980s with the enthusiasm around the
Single Market Program and the new faith in the future of integration. The
turn in research policy, however, preceded this general change of atmosphere; the decisions on ESPRIT and Framework Program were reached
when significant tensions were still tearing the Community apart.

Around 1980, the European Commission started to weight the possibilities to give new impulse for Community's activities research.[12] A crucial
component of these initiatives was the multi-annual framework program,
which would gather all the EC's research activities under a single institutional umbrella. It is, however, interesting that the first impulse for the
framework program came from the Council, frustrated with the inefficiency
of the Community's decision-making in the field. In its meeting of October
21, 1979, the Council expressed concern to improve the efficiency of the
development and execution of the Community's research programs.[13] It
asked the Commission to study the possibility of "setting Community indirect and concerted action programmes in the context of an appropriate
multi-annual framework" as well as to consider the "rationalization of
structures for the preparation, examination and implementation of
Community R&D programmes." The document produced by the
Commission services in response to the Council's request decidedly turned
down the idea of a single, multi-annual program: "The Commission has
considered this approach most carefully but has been forced to the

conclusion that an approach of this kind would be neither practicable nor desirable for the totality of indirect and concerted actions." The reasons for this conclusion were mostly administrative, including worries about an eventual administrative overload and the possibility to adjust the framework program with national programs. Instead, the Commission envisaged a "multi-program approach," grouping "some of the indirect and concerted programs by field of activity."[14] Commenting a Commission paper in May 1980, an official from DG XII (General directorate research) noted:

> The impression that many of our colleagues in the Member States will gain from the paper is that the Commission is saying we have looked at all these questions, we do not know what all the fuss is about, we think our present procedures and practices are excellent and we do not propose to make any changes.... If, in addition to giving this impression, we were to ignore their request for a report on the possibility of a multiannual framework programme for direct and indirect actions, we could be criticized strongly.

The fear was justified, because the Council was not alone asking for a study on the framework program: "We know that the officials with whom we have day-to-day dealings in CREST are strongly of this view."[15]

This is a crucial point. Over the years, the Commission had surrounded itself with a complex institutional web for consultation in the policy formulation phase as well as for translation of general ideas into concrete proposals for the Council and the Parliament. In research alone, there were several permanent bodies, including CREST, supported by a number of subcommittees,[16] CERD (Comité européen pour la recherche et développement) and CORDI (Comité de la recherche et du développement industriels). CREST was established in 1974 to play a triple role in counseling the Community institutions, coordinating the national policies of the member states, and consulting activities with the third countries. It was composed of high national officials and the representatives of the Commission. Created in 1972, CERD brought together independent experts with a background either in the academia, research administration, or industry. In October 1980, CERD organized a major conference, "1980-1990: Un nouveau développement de la politique scientifique européenne" [1980–1990: New Development of European Science Policy]. This two-day event brought together 150 experts (researchers, members of parliament, representatives of industry and labor organizations, public officials) to discuss the options and new dimensions of a common research policy.[17] CORDI, on the other hand, advised the

Commission since 1978 on issues related to research and industry. The members of the CORDI came from European interest groups such as UNICE (Union des industries de la Communaute européenne), CEEP (Centre européen de l'entreprise publique), FEICRO (Federation of European industrial cooperative research organizations), and CES (Conféderation européenne des syndicats).[18]

The existence of this patchwork of expert groups had a very clear rationale: compared with its responsibilities, the Commission was a relatively small institution that lacked expertise in many fields. Liaising with the scientific community and representatives of national institutions responsible for research policy offered a major resource for negotiations with governments. In addition, this vast and well-established machinery came handy not only in canvassing views and mobilizing expertise in the member states, but also in blurring the border between Brussels and the national capitals. The discussions on research policy support Peter Ludlow's argument that "conflictual model of Commission–Council relations is…totally misleading."[19]

It is in these continuous exchanges between the Commission and the various expert groups that the concept of the framework program was carefully shaped. At first it seemed as if the Commission's view of several sectorial framework programs would prevail. In the two CREST meetings held in September and October 1980, a majority of delegations agreed with the impossibility to realize a global framework program in the prevailing circumstances. Instead, the Commission's idea found wide support.[20] But in September 1981, a DG XII document concluded that a sectorial approach was insufficient to face the current challenges. "After successfully developing sectoral actions… the Community would have an interest, in the present conjecture, in focusing its initiatives on objectives (and no longer on priority sectors)."[21] This stance was translated into a major policy document, outlining the objectives of the Community's research effort for the next decade. Now the Commission was calling for an overall strategy: "Whatever the value and effectiveness of the programmes and coordination which the Community has carried out to date in the field of research and technology, it would seem that they are no longer adequate to make a sufficient response to the challenges which confront the Community or to rally national efforts in the light of that response."[22]

This new approach reflected a change of Commission in 1981. Speaking with the British ministers in London in September, Etienne Davignon,

Commissioner responsible for energy and industry, told he had been surprised at the lack of overall relationship between the Community research programs and policies as well as the Community's activities in general. Instead of policy, the discussion within the Commission had been focused on how to distribute money. This is something the ambitious member of the new Commission wanted to change.[23] Consequently, during the spring and fall of 1981, wary of having the governments' blessing for its initiatives, the Commission conducted a number of discussions with the national research policy administrations.[24] Together with the regular and confidential exchanges with CREST,[25] this gave the Brussels administration a relatively robust idea of what was possible to achieve.[26]

In October 1981, the Commission described the new framework program with the following words:

> What the Commission intends to develop, is an overall framework programme embracing all Community research, setting out against the options put forward for the Community as a whole, those actions and initiatives which are already being undertaken on the basis of the three treaties and those which are likely to be carried out in the future.[27]

The framework program would enable the Commission to discuss the policies of the member states and bring them together, rearrange priorities, and decide on the selection of joint actions and initiatives. The Commission also suggested doubling the Community's resources devoted to research by 1986.[28] The program was a major institutional innovation, which would ease decision-making in research along the lines initially envisioned by the Council: "For the Commission, the framework programme marks a fundamental shift from the sectorial approach hitherto, whereby individual R&D programmes have been approved on an ad hoc somewhat piecemeal basis, to a strategic approach whereby individual actions will be decided in relation to a previously defined, and annually updated, global strategy."[29]

The proposal for a framework program and the Community's overall strategy in research were discussed at the Council meeting of research ministers on November 9, 1981. The short minutes of the reunion reveal a broad consensus on the need to assume a global strategy for research and adopt a Community-wide framework program.[30] A few weeks later, the European Council adhered to the Commission's initiative to introduce a global R&D strategy for the 1980s.[31]

Encouraged by this positive reception, the Commission continued its consultations with CREST[32] and CORDI,[33] and a group of "Special Councilors,"[34] and produced the first major policy document on the framework program,[35] which ministers then discussed on March 8, 1982.[36] The reception was favorable, and the Commission was asked to submit a first outline of the program for the next Council meeting in June. Now, the Brussels executive was really pushing the issue. Writing to the Danish Ambassador Gunnar Riebenholdt on the eve of the Danish Council presidency in the second half of 1982, Etienne Davignon urged the Danes to speed up the decision-making process in the Council: "The Commission is hoping that Council will start its deliberation on all the proposals as rapidly as possible," he emphasized.[37] But there was not so much that the Danes could do. In June, the Council asked the Commission to present the final version of the framework program by November 1982,[38] but it took over a year until the document was approved.[39] The final decision was eased by an agreement at the higher political level a few days earlier, at the Stuttgart Summit of June 18, 1983.[40]

Research was in fact one of the new policies that the EC promoted in order to mask the deep-rooted political problems of the Community. On the eve of the dramatic 1983 Athens Summit, which *The New York Times* described as "the gravest crisis in the trade bloc's 26-year history,"[41] the governments of the ten member countries were keen to come up with a new agenda showing a continuing progress of integration. "All governments think that agreement on new policies would dress up with a bit of seasonal tinsel an EEC, which needs a better image," *Financial Times* reported on November 30, 1983, just a few days before the summit. In particular France and the Benelux countries, worried about the Community's minimal reaction to Europe's industrial and technological regeneration, and suspicious that the British and German would try to eject financial discipline into the Community, wanted to make research a central item on this agenda.[42]

While the Council approved the scientific goals of the framework program and the criteria for setting up corresponding programs, it failed to decide on the Commission's proposal for budget.[43] This was a major hiccup: one and a half years elapsed until the Council reached the final agreement.[44] Meanwhile, the Commission was forced to scale down its ambitions and postpone programs planned for 1985.[45] The main reason for this delay was the "budgetary constraints"[46] that continued to haunt the Community throughout 1983 and 1984. Once again, research policy became hostage to the bitter disagreements over the Community's finances, which this time mostly revolved around the question about the British rebate. Since some governments wanted to have a decision on the overall budget before discussing budgets for specific sectors,[47] any immediate progress on the issue started to look very unlikely. The framework program was associated with the budgetary dispute in two ways: in addition to the issue of timing, there was fierce opposition to all augmentation of funds for Community R&D.[48] For the most part, this opposition came from the two net contributors to the EC's budget, Britain and Germany.[49]

That Britain assumed a role as *l'Enfant terrible* in the discussion on the first framework program, is not very surprising. Since Britain's entry to the EC in 1973, the relations between London and its continental partners had been strained, and the return of a Conservative government in 1979 changed little.[50] The new Prime Minister Margaret Thatcher soon declared that she "cannot play Sister Bountiful to the Community."[51] Consequently, she was determined to fight for the British interest and advocated an uncompromising policy line that soon brought the Community on the brink of a total deadlock.[52] By 1983 these tensions had reached their low point. Thatcherite policies and the lingering disagreement over Britain's budgetary contribution had strengthened the domestic *Anti–Marketeers* and led to an increasing questioning about the desirability of Britain's continued engagement with Europe.[53] In the June 1983 Stuttgart European Council, Thatcher announced that she would not agree to resolution of other issues until the budgetary problem had been permanently solved.[54] This was not a constructive environment for discussing new openings, in particular those requiring an increase in the Community's resources.

Yet this is not to say that the British did not share the European concerns of weakening competitiveness and thus the rationale for European research cooperation. As a matter of fact, in Britain, the situation looked even more alarming: already by the mid-1970s Britain had fallen far behind Germany, France, and the Benelux countries in terms of productivity and the technological sophistication of exports. In a world of fierce economic concurrence, the country appeared badly prepared. "Because of its position at the lower end of the product sophistication spectrum, i.e. increasingly producing more mature products where price than non-price factors are important," Daniel T. Jones wrote in 1980, "Britain is likely to be one of the countries most severely affected by competition from the newly industrialized countries."[55] This realization led the British government to make concrete efforts to boost the country's performance, in particular in IT industries. In 1982, Kenneth Baker, Britain's new minister of state for information technology, announced a "Year of Information Technology," and in the following year, the government approved a major state-sponsored program of cooperative R&D. With a budget of total GBP 350 million, the Alvey Program, named after John Alvey, technical director of British Telecom and the chair of the committee behind the initiative, outsized all Britain's previous civil R&D programs. Yet, the conservative government remained torn between the need to support R&D and the anti-interventionist ideology.[56] Budgetary restraint continued to be dear for Prime Minster Thatcher, both for political but also ideological reasons.[57]

The other member state critical of augmenting the EC's research expenditure was Germany. The Federal Republic was the biggest contributor to the Community budget, and it was growing tired of paying the expanding bills.[58] For Bonn, any increase in the EC finances was highly undesirable as long as the government was forced to curb expenses at home. Crucially, this belt-tightening policy also affected research. Germany's Research Minister Andreas von Bülow was, however, generally favorable of the Commission's urge to start a new stage of a common research policy. In an article, published in the EC Magazine in May 1982, he said the efforts of the Brussels institution were of "great significance," especially in the time of putative or real "Europa-Müdigkeit" (Europe tiredness) and the preference of national interests in many policy fields.[59] As a matter of fact, the government body most reserved toward the EC's research policy initiatives was not research ministry but the ministry of finance, which feared that the new procedures envisaged for research could be stretched to the other sectors of the Community.[60] Unfortunately for the supporters of Community research policy, this attitude seemed to gain the upper hand in the Community affairs. In September 1981, the ministry argued that the EC effort in the field of research should be limited to sectors mentioned by the treaties or already recognized by the Council (energy, agriculture, environmental protection, raw materials, and nuclear safety): "In the industry-related cutting-edge technologies (telematics, biotechnology), selected as key areas in the Community Research Policy Committee, national industrial interests diverge strongly and hamper fertile Community activity. This will hardly be overcome in the near future."[61] No wonder the French delegation to Bonn saw a reason to complain about Germany's lame commitment to the EC: "Judging by their public statements or their confessions to newspapers, German officials seem quite willing to participate in a revival of Europe, as long as it would not cost them a Pfenning."[62]

The government change in October 1982 changed little. Although the new Chancellor Helmut Kohl announced he would make the EC one of the priorities of his government,[63] and giving a new impulse for European integration in the field of research and technology became one of the declared goals of the German Council presidency in the first half of 1983, the new leaders in Bonn appeared no more generous than their predecessors.[64] The new Research Minister Heinz Riesenhuber assumed a proactive and pragmatic approach to the Community and already in January 1983 presented a set of criteria for activity at the EC level. The "Riesenhumber criteria" ended up in the first framework program[65] and offered the EC an important guideline for the following years. But Germany's European drive was ambivalent,[66] and on research, the tensions between the foreign ministry and finance ministry prevailed. While the former, led by Hans-Dietrich Genscher, advocated for European cooperation in science and technology, the latter questioned the value of European collaborative programs, the framework program included. The ministry of research stood somewhere between, supporting collaboration in basic research, and approving the major components of the framework program. However, the ministry advocated selectivity for research efforts in more commercially relevant technologies and preferred to reduce the budget for some other European programs.[67]

Of the big member states the most enthusiastic about expanding the EC's role in the research sector was France, where the new socialist government, elected in 1981, rejected austerity and envisaged an ambitious increase in research spending. Between 1968 and 1981, research spending in France had fallen from 2.2 percent of the GDP to 1.8 percent. The goal of Research Minister Pierre Chèvenement was to reverse this trend by increasing the research budget to 2.5 percent by 1985.[68] According to *New Scientist*, describing the policy change as a "new love affair with research,"[69] Chevènement's main goal was to "implement what he calls Keynesian research policies" to rescue the country from economic distress.[70] The effort was extraordinary: "Many outsiders," the magazine observed, "are impressed by the hard cash and political commitment promised by the new government."[71]

Initially, the government's initiatives were firmly anchored in a national context and lacked a European dimension.[72] This reflected the reserved attitudes of the Elysée toward the EC, judged as ideologically alien to its neo-Keynesian project.[73] Only with the failure of President François Mitterrand's national experiment around 1983 did the government's focus change in favor of greater European cooperation. The Elysée's enthusiasm to develop new activity within the Community framework soon became associated with Laurent Fabius who followed Chevènement as the research minister.[74] Chairing the Council of Research Ministers during the French Council presidency in the first half of 1984, Fabius appeared as a firm supporter of the Commission's R&D initiatives. France's pro-European research policy continued even after Fabius became the prime minister and was followed by Hubert Curien. A current president of the European Science Foundation, and a former chairman of the council of ESA, Curien was "a committed and experienced Europeanist on technology matters."[75] However, France's record in championing for EC research remained mixed, and Paris never overcame its distrust of Brussel's leadership. Ultimately, Elysée's European endeavor of 1983 and 1984 culminated in the proposal for EUREKA, Mitterrand's grand plan for boosting European high-tech cooperation. EUREKA was an intergovernmental initiative that left little room for the Community.[76] The endeavor had a cool reception in Brussels, where it was seen as a potential competitor to the Commission's own plans.[77]

The discussions of the Framework Program coincided with another major Commission initiative to boost Europe's technological power. In late 1981 Etienne Davignon invited the directors of the twelve largest European IT companies to discuss the future of European industry. By early 1982, these roundtable discussions had already led to a work plan for a collaborative R&D program, whose pilot phase, launched a year later, received a favorable response: the call for proposal attracted over 200 applications, of which only 38 could be funded. In February 1984, this initial success of ESPRIT, the European Strategic Programme for Research and Development in Information Technology, led the Council of Ministers to approve the first phase of the ESPRIT program, carrying a budget of 1.5 billion ECU, half of which came from the EC and half from project participants.[78]

Wayne Sandholtz offers several explanations for ESPRIT's success: not only had the purely national telematics policies failed and made national politicians more favorable for cooperation, but also the major industrial actors were interested in the initiative, and an active Commission managed to mobilize a political coalition necessary to push the project off the ground. Davignon's strategy was to first sell the Commission's arguments to industry and then, with the support of the powerful business coalition, convince the national governments.[79] In addition, Davignon showed considerable policy leadership. In January 1984, *Science* magazine wrote that "Davignon's techniques for achieving his objectives have ranged from quiet diplomacy to pen arm-twisting." For example, when France, in a Research Council meeting in October 1983 advocated cutting ESPRIT's budget in half, Davignon declared the French position as "astonishing" and provoked a major news story in *Le Monde*, highlighting the discrepancy between this approach and the statements of President François Mitterrand to enhance European effort in the field of science and technology. By the next meeting, the French had modified their stance.[80]

But also, the time was ripe for a joint European effort. In the early 1980s, several European companies started to seek ways to better cooperate in the field of science and technology. In 1983, three of Europe's largest mainframe computer manufacturers—Siemens, International Computers Limited, and Bull—based in West Germany, Great Britain, and France, respectively, agreed to establish a new research center in Munich, focusing on artificial intelligence and "expert systems." According to *Science*, "[t]he new Munich center demonstrates how some companies are already building up a network of bilateral and multilateral research agreements independent of the EEC Commission's plans in case a failure to solve the general financial crisis within the Community means that ESPRIT has to be aborted."[81] Further, the success of ESPRIT showed the value of joint European effort and helped the member states to swallow the costs of the first framework program:[82] the 3750 million ECU budget of the program, intended for four years, raised the annual Community spending on research by 60 percent compared with 1982.[83]

The new determination to expand the EC's activities in the field of research and the creation and approval of the first framework program in 1980–1983/1984 marks a turning point in the efforts to promote integration in an increasingly vital policy domain. Several factors triggered this change: technological development, transformation of global economy, the Commission's policy leadership, the mobilization of industry, the success of ESPRIT, and finally, the gradual convergence of national positions. But the story remains incomplete without the important ideological and institutional continuity, which ultimately enabled the supporters of a more ambitious Community research policy to turn their initiatives into a compelling political case. This group of people would not have been as efficient in pushing through its agenda had it not been able to rely on previous practices, discussions, and above all, consolidated beliefs of science policy.

The main argument for a common research policy in the early 1980s is uncannily similar to the discussions two decades earlier: a relative weakness of European R&D, which was seen to stem, among other things, from the lack of coordination and cooperation between the European countries.[84] Usually, this weakness was articulated in terms of "retard" or "delay" in relation to Europe's two main economic competitors, the United States and Japan. In the Commission documents from the early 1980s, the notion of Europe's retard was omnipresent. Even the annual report for the Community's activities in 1983 devoted an entire page to a figure portraying the EC's retard in R&D spending in relation to the United States and Japan.[85] As it was in the 1960s, the image of the gap, coupled with notions such as "urgency" and "scarcity of time," became a great mobilizing force. For example, in August 1981, the Commission spoke about an "*urgent* need to bring together research efforts in the Community member states in a concerted whole as is the case already in the United States and Japan."[86] Another recurrent theme in the debate was the threat to Europe's independence by a lack of a strong R&D effort. "The autonomy of Europe, the demands of our society, the needs of the economy and industry as well as the aspirations of the scientific community all call for a true Community R&D strategy," a Commission document concluded in the same year.[87] To rally its cause, the Brussels institution was not afraid to paint a dystrophic view of the future: if Europe was to "continue as a set of 10 nation states operating independently, with a degree of voluntary cooperation from time to time across the frontiers when it suits particular governments at particular times," then "Europe's industries will be picked off one by one" and "[i]t will not be long before Europe is in the same position vis-à-vis the US and Japan as the less developed countries are to Europe – a source of relatively labor intensive industrial products and a market for the highly industrialized and profitable products of the more developed countries."[88] All these elements had been present in the previous debates two decades earlier.

The discourse was not limited to the European Commission. Even the British shared the concern about the gap and urged Europeans to act together. A discussion paper that the British government submitted at the Fontainebleau European Council of June 25–26, 1984, explicitly mentioned the challenge of the gap: "If the problems of growth, outdated industrial structures and unemployment which affects us all are to be tackled effectively, we must create the genuine common market in goods and services which is envisaged in the Treaty of Rome and will be crucial to our ability to meet the US and Japanese technological challenge."[89] Mr. Kenneth Baker, the British minister for Information Technology, saw ESPRIT as "an important element in ensuring the gap is closed."[90] The gap was also a major driving force in German high-tech policy. According to the German government, the new *Informationstechnik* (information technology) program, announced in 1984 and involving a budget DM 3 billion (1.14 billion in 1984) for five years, was a necessary response to the technological challenge of the United States and Japan.[91] Importantly, the German observations of the gap did not define domestic decisions only. In the ministry of foreign affairs, a common European high-tech policy was vigorously defended in response to American competition. The French debate of the early 1980s followed very similar lines. A memorandum calling for an ambitious EC industrial policy and prepared in September 1983 by Jacques Delors and other high-ranking officials, was predicted on the need of reacting to the high-tech challenge by United States and Japan.[92]

As we already know, the gap discussion fundamentally drew on the idea that research would generate economic growth and improve economic competitiveness. And in the early 1980s, the EC member states, Sandholtz writes, "believed more firmly than ever in the importance of IT in driving economic development and growth."[93] In the Community documents, one can observe an increasing emphasis on activities likely to boost industrial competitiveness. The five priority areas, established by the Council in December 1979 and reaffirmed in the subsequent Commission documents in 1980 and 1981, were energy, raw materials, agriculture, environment, and "certain industrial technologies." These sectors took up to 93.5 percent of the Community's funding.[94] A few years later, industrial technology had gained a more central place in the Commission's plans. The financial indications by objective that the Commission proposed in 1982 for the first framework program are illustrative: the greatest chunk (47.2 percent of total funding) was still going to energy research (developing nuclear fission energy, controlled thermonuclear fusion, developing renewable energy sources, rational use of energy),[95] but it was followed by "promoting industrial competitiveness" (28.2 percent of total funding, including removing and reducing barriers, new techniques and products for the traditional industries, new technologies). The remaining categories were of much smaller scale: "improving living and working conditions"

(10.3 percent), "stepping up development aid" (4.0 percent), "promoting agricultural competitiveness" (3.5 percent), "improving the effectiveness of the Community's scientific and technical potential" (2.3 percent), "horizontal action" (2.4 percent), and "improving the management of raw materials" (2.1 percent).[96] These same objectives and figures ended up in the final framework program.[97] Two years later, industrial competitiveness had gained even stronger emphasis: the Commission, forced to reconsider its priorities due to budgetary constraints, envisaged it as the only area with a relative increase of resources (36.8 percent compared to the 28.2).[98] In part, the increasing commitment industrial competitiveness stemmed from observations of technological change in certain "key" sectors. Perhaps the most fundamental difference to the 1960s was the IT revolution with its pervasive impact on industrialized societies. But advances in trade liberalization also served to sharpen international competition and to amplify concerns of competitiveness that ultimately would serve growth.

It is interesting that instead of a constraint, the supporters of a common research policy saw the tight economic situation of the early 1980s as an additional argument for increasing activity at a European level.[99] In the time of the budgetary constraint and rising cost of scientific research, the prospective of pooling resources and gaining extra guarantees of effectiveness and continuity at a Community level seemed to make sense.[100] A Commission report from May 1980 put this very clearly, stating that it was not

> possible to devise a new model for society, to secure the Community's political and economic autonomy or to guarantee competitiveness without a complete mastery of the most sophisticated technologies. Accumulated delays, rising costs and the fact that the necessary effort is too great to any member state to make individually are all compromising the Community's ability to compete in science and technology. The need for a proper response is all the more urgent: The Community dimension makes it possible to provide this response efficiently at minimum cost.[101]

Another Commission document pictured R&D as "one of the few sources of hope for the regeneration of Europe's economies and the solution of the many social problems the member states face."[102]

The altered economic and social circumstances also accounted for a greater emphasis on social aspects of research. Now, fighting unemployment and mastering the relationship between technological progress and social change occupied a central place in the Commission's proposals. "The move towards a new world energy order, the battle against inflation and unemployment, the problems of coming to terms with modernization and change, call for just as much innovation in social sphere as in the technological,"[103] the institution declared in 1981. Material anxiety and social unrest stemming from unemployment, as well as the possibility that European industries were facing extinction and its inhabitants relegating to "second class citizens of the world,"[104] occupied the minds of the architects of a common research policy and served to fuel the concerns about the gap.

Crucially, the Brussels institution continued to regard research policy as an engine for growth and competitiveness. A Commission document of February 1981 shows this continuity:

> More than ever, R&D remains a determining factor for economic growth, albeit moderate, and for social progress. R&D, which is at the root of innovation, is a prerequisite for the development and competitiveness of European industry.[105]

The framework program explicitly referred to the Article 2 of the EEC treaty assigning the Community the task of promoting harmonious development of economic activities, a continuous and balanced expansion, and an accelerated rising of standard of living.[106] The research policy debate of the early 1980s, thus, was firmly based on the premises laid out in the mid-1960s.

The evolution of the EC research policy over these two decades reveals a substantial character of European integration: over the years, and often in periods between major intergovernmental bargains, the Community moved into policy areas that were not specifically sanctioned by the treaties. This process was incrementally and systematically carried out by a devoted group of actors, relying on specific ideational frames. This institutionalization of new discourses, beliefs, and practices was supported by Article 235 of the EEC treaty that enabled the Community, by unanimous decision in the Council, to start activities in new areas if those were conceived necessary for achieving the aims of the Community.[107] Since one of these aims was economic prosperity, the supporters of a common research policy were able to formulate initiatives for concrete action. A good example here is the Council resolution of July 25, 1983, adopting the first framework program, which declared that "[w]hereas the Treaty establishing the European Community does not provide the specific powers of action required for the adoption of this resolution, [the Council] hereby adopts this resolution."[108] Even without a treaty basis, the Community was thus able to deepen its activities, gradually evolving into a significant political force within transforming Europe. When two years later the Single European Act, with its new Title VI (Article 130f to q), officially established Community research policy, it was more about reaffirmation of established practices than creating truly new ones.

Research policy was part of the "new agenda that emerged informally in the course of the 1970s and 1980s and was sanctioned by the Single European Act."[109] This agenda drew on the emerging consensus that joint measures were needed for facing the new societal and economic challenges in Western Europe. Moreover, the consensus was formed before Jacques Delors seized the initiative in 1985 as a new Commission president[110] and with the single market program spiraled the process of European integration into a new level. Research policy and the single market belonged to the same political package that toward the end of the decade led to a profound transformation of the European project.

NOTES

1. However, the total share of national funds going to international research cooperation had declined from 12.2 percent in 1967 to 8 percent in 1978. Commission working paper: Guidelines for the Common Policy for Science and Technology (1981–1986), July 1, 1981. BAC01111/88 (816), CAB.

2. Commission of the European Communities: Scientific and technical research and the European Community. Proposals for the 1980s. BAC0111/88 (816), CAB.

3. Une nouvelle phase du développement de la politique commune de recherche-développement (1981–1986). Document de travail, February 1981, BAC 0111/88 (813), CAB.

4. Bilan 1975–1980, recherche. Le 15 avril 1980. BAC01111/88 (812), CAB.

5. Une nouvelle phase de développement de la politique commune de recherche-développement (1981–1986). Internal document, November 10, 1981. BAC01111/88 (812). "During the 1970s and more especially the second part of the decade, the European Community experienced a profound evolution in political, economic and scientific matters. These

constitute a significant change of perspective compared with thinking prior to 1974, when the present Research and Development Policy was initiated." Research and Development—New Policy Document. 3rd version, September 9, 1981. BAC 01111/88 (818), CAB.

6. Commission of the European Communities: Scientific and technical research and the European Community. Proposals for the 1980s. BAC0111/88 (816), CAB.

7. *The Economist*, March 20, 1982.

8. Gaston Thorn, "Europe: To Be of Not to Be," Lecture at the European University Institute, May 24, 1984. *EUI Working Papers* 321, 8–9.

9. Of course, there was never a total standstill. Even in the "troubled years," a number of significant efforts were made to accelerate integration. These attempts, such as the Commission's early proposals for the internal market, the Genscher-Colombo—initiative of the German and Italian foreign ministers, which amounted to the Solemn Declaration on European Union in 1983 and the European Parliament declaration on European Union in 1984, whatever their immediate impact, paved the way for future decisions. Nicola Fielder, *Western European Integration in the 1980s: The Origins of the Single Market* (Bern and Berlin: Peter Lang 1997), 60–63; Immo Stabreit, "Die 'Feierliche Deklaration zur Europäischen Union'—eine Etappe auf dem Weg zu einem Vereinten Europa," *Europa-Archiv*, Folge 15 (1983), 445–452; Hans Stark, *Helmut Kohl, L'Allemagne et l'Europe. La politique d'intégration européenne de la République fédérale 1982–1998* (Paris: L'Harmattan, 2004), 31–35; also advances in monetary field were important: George Zis, "The European Monetary System 1979–1984: An Assessment," *JCMS*, 23 (1984): 45–72; Emmanuel Mourlon-Druol, *A Europe Made of Money: The Emergence of the European Monetary System* (Ithaca, N.Y.: Cornell University Press, 2012).

10. Stark, *Helmut Kohl, L'Allemagne et l'Europe*, 37.

11. Desmond Dinan, *Europe Recast. A History of European Union* (Houndsmills, Basingstoke: Palgrave Macmillan, 2004): 167–201.

12. Une nouvelle phase du développement de la politique commune de recherche-développement (1981–1986). Document de travail, February 1981, BAC 0111/88 (813), CAB; La recherche scientifique et le développement technologique (stratégie commune pour les années 1980), Eléments de réflexion en vue de la réunion du 21 septembre 1981 chez Monsieur le Vice-Président Etienne Davignon. Bruxelles, le 21 septembre 1981. BAC 01111/88 (818).

13. Note for attention of Mr. Schuster: Report on the further development of the Common Policy in Science and Technology. Brussels, May 21, 1980, BAC 0111/88, (808) CAB.

14. The Common policy in science and technology—priorities and organization. Report to the Council in response to its request of December 20, 1979, May 1980, BAC0111/88 (808), CAB.

15. Note for attention of Mr. Schuster: Report on the further development of the Common Policy in Science and Technology. Brussels, May 21, 1980, BAC 0111/88, CAB (808).

16. Principaux organes consultatif communautaires, le 14 janvier 1981. BAC 0111/88 (812), CAB.

17. Une nouvelle phase de la developpement de la politique commune de recherche-developpement (1981–1986). Document de travail, February 1981, BAC 0111/88 (813), CAB.

18. Recherche scientifique et développement technologique. Principaux organes consultatif communautaires, le 14 janvier 1981. BAC 0111/88 (812), CAB.

19. Peter Ludlow, "The European Commission," in *The New European Community. Decisionmaking and Institutional Change*, eds. Robert O. Keohane and Stanley Hoffmann (Boulder, San Francisco: Westview Press 1991), 102–104.

20. Avis du CREST sur la communication de la Commission au conseil "La politique commune dans le domaine de la S&T: priorités et l'organisation," le 5 novembre 1980. BAC0111/88 (810), CAB.

21. La recherche scientifique et le développement technologique. Stratégie commune pour les années 1982–1986, le 18 septembre 1981. BAC0111/88 (819), CAB. The author's translation from French.

22. Commission of the European Communities: Scientific and technical research and the European Community. Proposals for the 1980s. BAC0111/88 (816), CAB.

23. Note for the attention of Mr. Bouwers, Cabinet of Mr. Davignon: Mr. Davignon's meeting with British ministers in London, September 25, 1981. BAC0111/88 (819), CAB.

24. Compte rendu de mission: Entretien entre M. Davignon et M. Aigrain (Secrétaire d'Etat à la recherche—France). Paris, le 8 avril 1981. BAC0111/88 (816), CAB; Aktennotiz: Besprechung von Fragen der Europäischen Forschungs- und Technologiepolitik mit Vertretern der verschiedenen Deutschen Ministerien. Bonn, den 17. April 1981. BAC0111/88 (816), CAB; Note à l'attention de M. Schuster: "Lignes directrices," réunions à Londres. London, le 28 avril 1981. BAC0111/88 (816), CAB; Entretien avec le monsieur le ministre Chevènement, Paris, le 2 octobre 1981. BAC0111/88 (820), CAB; Entretien du 9 octobre entre M. Tesini, Ministre Italien de la recherché et M.M. Davignon et Fasella. BAC0111/88 (821), CAB; Note à l'attention de Monsieur E. Davignon, Vice-President de la Commission: Entretien avec le Monsieur le Ministre von Bülow, Bonn, le 20 octobre 1981, Bruxelles, le

21 octobre 1981. BAC0111/88 (822), CAB; Preparation of the Research Council—9 November 1981, Note of a meeting in London at 5 pm on Monday, 2 November. BAC0111/88 (823), CAB.

25. Note de dossier: Politique commune dans la domaine de la S/T (DG XII), le 27 mai 1981. BAC0111/88 (816), CAB.

26. Recherche scientifique et développement technologique. Principaux organes consultatif communautaires, le 14 janvier 1981. BAC 0111/88 (812), CAB.

27. Communication from the Commission to the Council: Scientific and technical research and the European Community. Proposals for the 1980s. COM(81) 547 (final), Brussels, October 12, 1981. BAC0111/88 (821), CAB.

28. Commission of the European Communities: Scientific and technical research and the European Community. Proposals for the 1980s. BAC0111/88 (816), CAB; Communication from the Commission to the Council: Scientific and technical research and the European Community. Proposals for the 1980s. COM(81) 547 (final), Brussels, October 12, 1981. BAC0111/88 (821), CAB.

29. Community R&D strategy for the 1980s—first outline of a framework program 1984–1987, guidance debate at Council. Brussels, June 10, 1982. BAC 107/1993 (379), CAB.

30. At the meeting, the Council invited the Commission to make concrete proposals for a general framework program, aiming at a global strategy for research; stimulating the efficiency of European the research system; the optimal use of available resources; orienting R&D programs toward the needs of industrial innovation; considering social consequences of technological development; improving the consultative mechanisms for the preparation and execution of Community research programs; and the importance of the continuing efforts to improve the diffusion of knowledge within the Community. Projet de procès-verbal sur la 736ème session de la Conseil tenu à Bruxelles le 9 novembre, 1981. BAC0111/88 (823), CAB.

31. Réunion planaire des "Conseillers spéciaux" en matière de Science et de Technologie, Bruxelles, le 8 janvier 1982. BAC 107/1993 (358), CAB.

32. At the CREST meeting on February 26, 1982, most delegations responded positively to the program, though considering that it should be a Community instrument, and in no case involved in the planning of national research policies. Synthèse des débats du CREST du 26.2.82. Bruxelles, le 3 mars 1982. BAC 107/1993 (378), CAB.

33. Programme cadre des activités scientifiques et techniques communautaires: principes et modalités. Bruxelles, le 10 février 1982. BAC 107/1993 (358), CAB.

34. Réunion planaire des "Conseillers spéciaux" en matière de Science et de Technologie, Bruxelles, le 8 janvier 1982. BAC 107/1993 (358), CAB. It is interesting that the General Director of DG XII, P. Fasella, contacted also major European companies. Letter from Karl Heiz Beckurts, Midglied des Vorstands der Siemens AG to P.Fasella, München, February 8, 1982. BAC 107/1993 (358), CAB.

35. Programme cadre des activités scientifiques et techniques communautaires: première esquisse. Bruxelles, le 2 février 1982. BAC 107/1993 (378), CAB.

36. Framework Programme for the scientific and technical activities of the Community. Principles and methods of working. Preparatory note for the Council on 8 March 1982. Brussels, February 23, 1982. (SEC(82)310). BAC 107/1993 (378), CAB; Projet de process-verbal de la 756ème session du Conseil, Bruxelles, le lundi 8 mars 1982. 107/1993 (358). The Council, convening on 8 March 1982, invited the Commission to make a new draft to be examined in the following Council meeting. The Council also approved the Commission's proposition to create an ad hoc group, composed of the representatives of the Commission, CREST and the personal representatives of ministers to study the procedures of preparation, decision and evaluation of programs. Communication à la Presse 5387/82 (presse 25), Archive of the European Council, Brussels (hereafter AEC).

37. Letter from Etienne Davignon to Gunnar Riberholdt, Brussels, May 5, 1982. BAC 107/1993 (358), CAB.

38. Extrait du procès-verbal de la 785ème session du Conseil du 30 juin 1982. BAC 107/1993 (358), CAB.

39. Commission of the European Communities, *Seventeenth General Report on the Activities of the European Communities 1983* (Brussels-Luxembourg 1983), 224–225.

40. Georges Saunier, "La genèse du premier programme-cadre européen. Un regard Français (1981–1984)," in *La construction d'un espace scientifique commun? La France, la RFA et l'Europe après le 'choc du Sputnik,'* eds. by Corine Defrance and Ulrich Pfeil (Brussels: P.I.E. Peter Lang, 2012), 82.

41. *The New York Times*, December 8, 1983.

42. *Financial Times*, November 30, 1983.

43. Commission of the European Communities, *Seventeenth General Report*, 224–225.

44. Commission of the European Communities: *Eighteenth General Report on the Activities of the European Communities*. Brussels-Luxembourg, 1984, 214.

45. Such as ecoclimatology, natural disasters, agronomics research, non-ionising radiation, the technology and science of foodstuffs, health and

safety at work. Commission of the European Communities, Communication to the Council: Community Research Priorities COM (84) 287 final, Brussels, May 24, 1984. BAC 107/1993 (360), CAB.

46. Commission of the European Communities, *Eighteenth General Report*, 214.

47. Note à l'attention de Monsieur E. Davignon, Vice-Président de la Commission: Entretien avec le Monsieur le Ministre von Bülow, Bonn, le 20 octobre 1981, Bruxelles, le 21 octobre 1981. BAC0111/88 (822), CAB.

48. Entretien avec le Monsieur le Ministre von Bülow, Bonn 20 octobre 1981. Bruxelles, le 15 octobre 1981. BAC0111/88 (822), CAB.

49. Préparation du Conseil de recherche du 29 juin 1984, Réunion du groupe "Recherche" du 18 juin 1984. BAC 107/1993, CAB; Conseil recherche du 29 juin 1984, débat sur les priorités: rappel des positions prises par le conseil et ses instances sur le programme cadre des activités S/T communautaires (depuis le 15 juillet 1983). CAB 107/1993 (361), BAC.

50. For Britain's European policy in the 1970s, see David Gowland and Arthur Turner, *Reluctant Europeans: Britain and European Integration, 1945–1998* (Essex: Pearson Education, 2000), 169–229.

51. Gowland and Turner, *Reluctant Europeans*, 254.

52. Roy Denman, *Missed Chances, Britain and Europe in the Twentieth Century* (London: Cassell 1995), 243–261; Gowland and Turner, *Reluctant Europeans*, 244–258.

53. Maureen Murr, *Britain and Europe During 1983. A Bibliographical Guide* (Sussex: Harvester Press Microfilm Publications, 1984), 7.

54. Gowland and Turner, *Reluctant Europeans*, 256.

55. Daniel T. Jones, "British Industrial Regeneration: The European Dimension," in *Britain in Europe*, ed. William Wallace (London: Royal Institute of International Affairs, 1980): 118, 122.

56. Wayne Sandholtz, *High-Tech Europe: the Politics of International Cooperation* (Berkley and Los Angeles: University of California Press, 1992): 155–156.

57. Gowland and Turner, *Reluctant Europeans*, 261.

58. Stark, *Helmut Kohl, L'Allemagne et l'Europe*, 26–27.

59. *EG Magazin*, May 1982.

60. Note à l'attention de Monsieur J.C. Brouwers, Cabinet du Vice-Président Davignon. Entretien du Vice-Président Davignon avec le Ministre v. Bülow. Bruxelles, le 9 juin 1982. BAC 107/1993 (358), CAB.

61. 13. Sitzung des Bundeausschusses für Forschung und Technologie; Bericht der EG-Kommission zu dem Mandat von 30.5.1980 (Lösung der

Finanzprobleme der Gemeinschaft), Bonn, den 10. September 1981, 421.46 (123525), PA AA. The author's translation from German.

62. La République Fédérale et la Relance de l'Europe. Cable from the French delegation in Bonn to Paris, September 4, 1981, 1930INVA/4971, MAEF. The author's translation from French.

63. Stark, *Helmut Kohl, L'Allemagne et l'Europe*, 25–33.

64. Zukunftsaspekte Europäischer Forschungspolitik. Pressartikel von Dr. Heinz Riesenhuber, Bundesminister für Forschung und Technologie, Bonn 1983. BAC 107/1993 (358), CAB.

65. The Framework Program states that "Community activity can be justified in the following cases: research on a very large scale for which the individual Member States could not, or could only with difficulty, provide the necessary finance and personnel; research, the joint execution of which would offer obvious financial benefits, even after taking account of the extra costs inherent in all international cooperation; research which, because of the complementary nature of work being done nationally in part of a given field, enables significant results being obtained in the Community as a whole for the case of problems whose solution requires research on a large scale, particularly geographical; research which helps to strengthen the cohesion of the common market and to unify the scientific and technical area and research leading, where the need is felt, to the establishment of uniform standards." Commission of the European Communities. Framework Programme for Research 1984–1987, COM (83) 260, final, Brussels, May 17, 1983.

66. Stark, *Helmut Kohl, L'Allemagne et l'Europe*, 25–27.

67. Sandholtz, *High-Tech Europe*, 154.

68. *New Scientist*, July 9, 1981.

69. *New Scientist*, January 7, 1982.

70. *New Scientist*, July 9, 1981; Saunier 2012, 86.

71. In reality, the break with the past was only partial: Mitterrand's predecessor Valery Giscard d'Estaing had already turned the political spotlight on research and planned to increase national R&D spending to 2.3 percent of the GDP by 1988. "Mitterrand gives a French lesson." *New Scientist*, January 7, 1982.

72. Saunier, "La genèse du premier programme-cadre européen," 88–89.

73. Stark, *Helmut Kohl, L'Allemagne et l'Europe*, 37.

74. Saunier, "La genèse du premier programme-cadre européen," 89, 91.

75. Sandholtz, *High-Tech Europe*, 151–152.

76. On Eureka, see: Georges Saunier, "Eurêka: un projet industriel pour l'Europe, une réponse à un défi stratégique." *Journal of European Integration History*, 12 (2006): 57–74; John Peterson, *High Technology and the Competition State: An Analysis of the EUREKA Initiative* (London: Routledge & Kegan Paul, 1993); Philippe Braillard and Alain

Demant, *Eureka et l'Europe technologique* (Brussels: Bruylant, 1991); Veera Mitzner, "Almost in Europe? How Finland's Embarrassing Entry into Eureka Captured Policy Change," *Contemporary European History* 25/3 (2016): 481–504.

77. David Dickson, "France Seeks Joint European Research," *Science*, May 10, 1985.

78. Sandholtz, *High-Tech Europe*, 4.

79. Ibid. 159–165.

80. David Dickson, "Europe Seeks Joint Computer Research Effort," *Science*, January 6, 1984.

81. Ibid.

82. Herbert Allegeier, interview by Arthe Van Laer, Brussels, November 4, 2010. In HISTCOM.2 Histoire interne de la Commission européenne 1973–1986.

83. Saunier, "La genèse du premier programme-cadre européen," 84; Arthe van Laer, "Research: Towards a New Common Policy," in *The European Commission 1973—86. History and memoires of an institution*, eds. Eric Bussière et al. (Brussels-Luxembourg: European Commission, 2014), 288.

84. It is interesting, though, that the lack of manpower or resources was no longer identified as a source for Europe's deficiencies. The weakness was of structural origin. Commission of the European Communities: Scientific and technical research and the European Community. Proposals for the 1980s. BAC0111/88 (816), CAB.

85. Commission of the European Communities, *Seventieth General Report*, 227.

86. New policy document on research and development, August 21, 1981. BAC0111/88 (817), CAB.

87. Communication from the Commission to the Council: Scientific and technical research and the European Community. Proposals for the 1980s. COM(81) 547 (final), Brussels, October 12, 1981. BAC0111/88 (821), CAB.

88. New policy document on research and development, August 21, 1981. BAC0111/88 (817), CAB.

89. A discussion paper submitted by the British government at the Fontainebleau European Council of June 25–26, 1984. Quoted in: David Gowland and Arthur Turner, *Britain and European Integration 1945–1998. A Documentary History* (London and New York: Routledge, 2000), 173–175.

90. *Financial Times*, March 28, 1983.

91. Sandholtz, *High-Tech Europe*, 153.

92. Warlouzet, *Governing Europe*, 127–128.

93. Sandholtz, *High-Tech Europe*, 144.
94. The Common policy in science and technology—priorities and organization. Report to the Council in response to its request of December 20, 1979, May 1980, BAC0111/88 (808), CAB. New policy document on research and development, August 21, 1981, BAC0111/88 (817), CAB.
95. The great focus on energy research is largely explained by the legacy of Euratom—after all, activity under the auspices of the atomic energy community was the most important area of the EC's research from the beginning—but also by the political climate after the energy crisis in 1973–1974. To West Europeans, the sudden end of the supply of cheap energy came as a major shock that led to an intensive search for ways to increase the region's independence from the oil-producing countries. As late as in 1978 energy still swallowed almost 60 percent of the Community's research budget. Kurt-Jürgen Maaß: Die Forschungspolitik der Europäischen Gemeinschaft, Bonn, BAC0111/88 (808), CAB.
96. Commission of the European Communities, *Seventieth General Report*, 226.
97. Commission of the European Communities. Framework Programme for Research 1984–1987, COM (83) 260, final, Brussels, May 17, 1983.
98. Commission of the European Communities, Communication to the Council: Community Research Priorities COM (84) 287 final, Brussels, May 24, 1984. BAC 107/1993 (360), CAB.
99. Recherche scientifique et développement technologique. Principaux organes consultatif communautaires. Le 14 janvier 1981. BAC 0111/88 (812), CAB.
100. Commission of the European Communities: Scientific and technical research and the European Community. Proposals for the 1980s. BAC0111/88 (816), CAB; Compte rendu de l'entretien entre M. le President M. Davignon et les membres titulaires du CREST, le 14 juillet 1981. BAC0111/88 (816), CAB.
101. Quoted in: New policy document on research and development, August 21, 1981. BAC0111/88 (817), CAB.
102. Ibid.
103. Commission of the European Communities: Scientific and technical research and the European Community. Proposals for the 1980s. BAC0111/88 (816), CAB.
104. New policy document on research and development, August 21, 1981. BAC0111/88 (817), CAB.
105. Une nouvelle phase du développement de la politique commune de recherche-développement (1981–1986). Document de travail, February 1981, BAC 0111/88 (813), CAB. The author's translation from French.

106. Commission of the European Communities. Framework Programme for Research 1984–1987, COM (83) 260, final, Brussels, May 17, 1983.

107. "If any action by the Community appears necessary to achieve, in the functioning of the Common Market, one of the aims of the Community in cases where this Treaty has not provided for the requisite powers of action, the Council, acting by means of a unanimous vote on a proposal of the Commission and after the Assembly has been consulted, shall enact the appropriate provisions." *Treaty establishing the European Economic Community* (Rome, 25 March 1957).

108. Résolution du Conseil du 25 juillet 1983 relative à des programmes cadres pour des activités communautaires de recherche, de développement et de démonstration, et au premier programme cadre 1984–1984. Reproduced in Journal official des Communautés européennes, N° C 208. English translation in the framework program: Commission of the European Communities. Framework Programme for Research 1984–1987, COM (83) 260, final, Brussels, May 17, 1983.

109. Ludlow, "The European Commission," 88.

110. Already in March 1984 the Council, urging for the creation of a "real economic union," set the following objectives: the convergence of the economic policies and Community action able to promote productive investment, the development of the scientific and technologic potential of Europe, creation of an internal market where European companies can profit from the Community dimension, and the defense and promotion of employment. Projet de procès-verbal de la 922ème session du Conseil tenue à Bruxelles, le mardi 28 mars 1984. BAC 107/1993 (361), CAB. The single market didn't belong to the Community's top priorities, and it virtually disappeared from the agenda of the 1984 Fontainebleau summit. Lord Cockfield, *The European Union. Creating the Single Market* (London: Wiley Chancery Law, 1994), 23–24.

CHAPTER 10

Conclusion and Further Thoughts

This book has shown how a specific conception of research policy, based on the conviction that scientific research has economic benefit for Europe, formed the cornerstone of the European Community's and later the European Union's activity in the field. This view was of fundamental importance for the establishment of institutions and political initiatives in an area where the EC/EU initially lacked policy competence. Linked to the continuous political desire for growth, the increasingly pervasive imperative of competitiveness, as well as to popular language about Europe's scientific and technological retard vis-à-vis its main commercial competitors, it provided research of political and ideational justification as a distinct EC/EU-level policy domain.

As much as change, this book tells a story of continuity, longevity, and durability. It illustrates the power of past in the present. We have seen how the growth-centric conception of research policy was first developed beyond the Community structures and effectively distributed and consolidated by actors working within the realm of the OECD and some national institutions. Soon, the newly formed European Commission, which closely followed the work of the OECD and was keen to expand the Community's competences into new areas, adopted this utilitarian idea of research and made it to its own. The moment for a new Community policy serving growth seemed propitious: committed to securing the expansion of national economies in the aftermath of World War II and in the context of the Cold War competition, and concerned about an allegedly increasing

© The Author(s) 2020
V. Mitzner, *European Union Research Policy*, Europe in Transition:
The NYU European Studies Series,
https://doi.org/10.1007/978-3-030-41395-8_10

separation of Europe from the technologically more advanced United States, the European political and economic elites appeared open to arguments for unifying European resources in science and innovation. With the postwar market liberalization, rapid technological change, and the gradual shift of production to the newly industrialized countries, the case for joint European research effort seemed plausible.

Maria Nedeva and Linda Wedlin have divided the development of European Union research policy in two main phases: "Science in Europe" and "European Science." According to these two scholars, the first phase, "Science in Europe," which roughly covers the time period analyzed in this book, was characterized by the principle of subsidiarity—an idea that European-level action would only be appropriate if action at a national level was insufficient; focus on increasing collaboration amongst researchers from different European countries; and concerns of the technology gap, which Nedeva and Wedlin call the "European paradox," as Europe as a world leader of science seemed to lag behind in the industrial and economic exploitation of scientific ideas. Furthermore, "Science in Europe" witnessed three distinct waves: the establishment of large, transnational facilities, such as European Organization for Nuclear Research (CERN), Euratom, European Molecular Biology Organization (EMBO), and European Molecular Biology Laboratory (EMBL); the creation of "diffuse" organizations, such as European Cooperation in Science and Technology (COST) and the European Science Foundation (ESF), including all fields of research and providing platforms for cross-national cooperation rather than supporting science at European level; and finally the creation of the EC/EU Framework Program for research, bringing the scattered research programs under one administrative umbrella. For Nedeva and Wedlin, "Science in Europe" was a period "during which a partial and fragmented science and research system was developed." It complemented rather than challenged national science, research, and innovation arrangement; it had a limited impact on universities, research institutes, and national funding agencies and landscapes; and it "didn't in a rule, include explicitly European-level research performing organizations."[1]

This book doesn't challenge Nedeva's and Wedlin's analysis. For a long time, the EC/EU research policy had a partial character—the first initiatives were not designed to directly challenge activity in national settings. The proponents of European level activities in the field, while often ambitious and visionary, were aware of the political realities. Even the "European paradox" argument is solid, although the debate had its ebbs and flows and occurred in different variations depending on the context and advocates. Fears of European technological retard did push initiatives forward, at different degrees of success, giving them rationale and a sense of urgency. Also, European level research facilities were still few and far in between, and decisions in Brussels barely influenced the daily operations of universities and research institutes in the EC member states.

However, with a closer historical analysis, we also see an emerging policy that was constantly challenged, contested, and vetted by a number of groups, and in particular those who would mostly have been affected by it. And this is the vital "more to the story" that Nedeva's and Wedlin's and others' accounts do not fully tell. This book also unveils the bigger ambitions of policy-makers and dreamers by fletching out the protracted struggles for initiatives that either never quite made it to the fruition or were realized as truncated versions, to the disappointment of many. It sheds light to the already forgotten and fluid policy spaces where the feasible and unfeasible were defined, and where future pathways were sometimes painfully determined. Crucially, these spaces were shaped by broader political, social, and economic circumstances within and beyond Europe, and the prevalence and strength of specific ideational shifts and conditions, such as the fading of the naïve enthusiasm about science that had characterized the immediate postwar period; increasing uneasiness about growth policies and calls for better accountability of economic and scientific activity; new limitations imposed by economic austerity; persisting distrust in European level bureaucracy among academics and national policy-makers; and ongoing political battles in other policy areas within the EC context. Yet one distinct thread runs through all these events and debates: a basic agreement on the beneficial role of research and research policy in achieving greater growth and competitiveness, and the need to create joint political mechanisms to that end. How this would eventually be achieved remained open for contestation.

Besides being a story of continuity and struggle, this book also is an account of the vulnerability of a political idea that survived, grew big, and eventually transformed European research. It was just as strong as its proponents, which at the beginning were not many. Although the Commission made sure that science policy experts would be involved in the elaboration of the various schemes from the start, these experts often represented governments and national science funding bodies rather than the broader science community. The lack of active participation of scientists and their advocates, as well as of industry and the wider public in the early discussion on the topic, is striking; until the 1980s, most proposals were developed in technocratic milieus with little public attention. The backing of pro-European associations such as Union des industries de la Communauté européenne (UNICE), the European Parliament, and some individuals from national research institutions and administrations remained too weak and fragmented to put significant pressure on key decision-makers.

This is the tragedy of the early EC/EU research policy. Politicians in the EC member states, whose main concern was often winning the support of the electorate, tended to focus on issues with more immediate political benefit. It was not the case that the citizens in the Community member countries had opposed the idea of the Europeanization of scientific and technological activities: in the Community's surveys of public opinion, research has consistently been at the top of the list of policies that people think should not be managed exclusively at a national level.[2] Research was, nevertheless, not the kind of "high politics" that would be surrounded by glamour and drama and that would have reached the headlines. It remained distant to the lives of ordinary citizens who barely knew of the Commissions initiatives.[3] It is also exactly this technical, abstract, complex, and distant nature of this policy domain that at least in part explains the persisting gap between rhetoric and reality in European research policy from the mid-1960s to the mid-1980s. For national politicians, expressing catchy phrases and stirring rhetoric about science and its many promises for the future, could be politically advantageous but also relatively safe: sacrifices in this sector rarely had a direct impact on the lives of voters, and therefore, they seldom provoked significant public protests. The late 1960s' protests at the Ispra Euratom Joint Research Centre stand out as a notable exception, as multiple employees were threatened to lose their jobs as a result of decisions made in Brussels. But in most cases, lip service in research policy came at a low political price.

In the absence of a systematic backing of scientists or their representatives, the Commission alone was too weak to induce the national governments to truly commit to developing a Community research policy. Moreover, Commission's strategy was not always consistent: important tensions prevailed between the DGs, while each Commissioner in charge of research pursued a slightly different policy. The existence of DG Research, under various names since 1967, however, became a central structure for guaranteeing institutional and ideational continuity. The DG constituted a center where bureaucrats and experts, sharing some important ideas and interests, could further elaborate schemes for new Community activity. This finding supports Katia Seidel's research on DGs IV (agriculture) and VI (competition), which shows that the administrative cultures developed in those DGs and the subsequent socialization of officials into these cultures, "helped to bring about policies in relevant areas, and then to keep them on track."[4]

Through the discussions initiated by DG Research and others, the EC gradually formed an arena for "Europeanization"[5] in research where socialization of different experts and political actors could take place. Intra-European connections were reinforced, and similarities increased, while "Europe" was given new contents and contours. At the same time, it would be misleading to argue that all contacts within the realm of the research policy debate would automatically have led to greater cohesion. These exchanges also exposed to governments and experts the diverse character of European research systems and made them aware of the difficulty in creating a strongly integrated common policy. Yet they prepared the ground for subsequent initiatives that would take off with a more favorable political momentum.

In the early 1980s, the various plans and scattered activities of the previous twenty years finally evolved into something that could be said to have an important scope and political and economic significance. The first step into this new direction was the launch of sizeable technological programs, the pioneer being European Strategic Program on Research in Information Technology (ESPRIT), an initiative that was intended to develop a European strategic scheme based on collaboration between major European companies, small- and medium-sized firms, universities, and research institutes. The project was successful: of the proposals received in response to the first call, less than 25 percent could be funded. Not surprisingly, after 1985, ESPRIT became the model for successive Community programs, such as Research in Advanced Communications in Europe (RACE) and Basic Research in Industrial Technologies (BRITE).[6] The expansion of these activities led to the creation of the first framework program for research (1984–1987) in 1984. The €3.75 billion budget of the framework program, together with the budget for ESPRIT of €750 million, signified a threefold increase in the Community's research spending in 1982. Moreover, the program sparked off a steady and fast increase of the Community's research budgets. Peter Tindemans has observed that after ESPRIT, there were a growing number of scientists and industrialists involved in the consultations, projects, and evaluations, and thereby directly benefiting from the Community's activity in the field. This created "an almost unstoppable dynamic of pressure [that] arose in national capitals and in the Directorates-General responsible for the FP's [framework programs] to increase the budget."[7]

Although the role of the EC as a focal point of common European research activities remained contested, as was demonstrated by the creation of Eureka, an intergovernmental initiative intended to promote "near-market research," in 1985, and by the difficult negotiations of the subsequent framework programs, the political context had changed. National governments were now more willing to see the EC take a central role in the field, and a horizontal and vertical growth of Brussels-led activities ensued. All this was supported by the favorable conjuncture in European integration that ensued the single market program. The first framework programs were still focused on advancing applied, market-oriented research, but during the 1990s, these programs became more comprehensive and versatile as they included new mechanisms and stretched into different disciplines and sectors. They also opened up for countries outside the EC/EU framework, developed collaboration with other European and international organizations, and finally became more closely integrated with the EC/EU core policies.[8] An EU-centric European research space started to emerge.

This is, indeed, where Wedlin and Nedeva see the shift to "European Science," marked by a changing understanding of the "European added value." Whereas the previous EC/EU activities had been primarily focused on fostering collaboration between researchers, now there was a widely shared desire to increase the level of integration in different aspects of European science and research. New tools were designed to strengthen the European knowledge base and to support basic research through competitive funding.[9]

A key concept here is the European Research Area (ERA).[10] Originally dreamed up by Commissioner Ralf Dahrendorf in the 1970s, the idea was revived by Commissioner Alberto Ruberti in the 1990s, and finally pushed to the center of the European political agenda by Commissioner Philippe Busquin. Endorsed as an ambitious effort to pool European scientific and technological resources, the initiative marked a clear break with the distributive approach that had characterized the EU activities in the preceding years.[11] While the definition of what the ERA stands for has evolved over the years, it has brought new inclusiveness into the EU research policy and signaled a move toward a more Europeanized research policy field, "where the ERA agenda provides justification for the adoption of new institutions and funding tools."[12] ERA aimed to increase the Commission's autonomy in initiating projects and programs with direct implications for national activities.[13] In 2012, the Commission defined ERA as "a unified research area open to the world based on the Internal Market, in which researchers, scientific knowledge and technology circulate freely and through which the Union and its Member States strengthen their scientific and technological bases, their competitiveness and their capacity to collectively address grand challenges."[14]

ERA was launched in the political framework of the Lisbon European Council of March 2000, where the EU set itself the new strategic goal of becoming "the most competitive and dynamic knowledge-based economy in the world capable of sustainable economic growth, with more and better jobs and greater social cohesion."[15] In Lisbon, research and development were drawn to the center of the EU's strategy for achieving its goal by 2010. Essentially, scientific research, technological development, and innovation were defined as key factors in growth, competitiveness, and employment in a knowledge-based European economy.[16] This trend was continued in the EU's new growth strategy, Europe 2020, launched in 2010. One of the seven flagship initiatives announced in the strategy was Innovation Union, aiming to "improve conditions and access to finance and innovation, to ensure that innovative ideas can be turned into products and services that create growth and jobs."[17] ERA was considered a core element of the Innovation Union and the Europe 2020 strategy.[18] The EU's 2002 Barcelona target to achieve R&D funding of 3 percent of GDP—an objective largely motivated by the higher R&D investment levels in the United States and Japan—gave ERA additional political support and visibility.[19]

According to Wedlin and Nedeva, "European Science" has created "an organizational space for the support of research that is aligned, and in competition with national-level research spaces."[20] A key invention in this space was the European Research Council (ERC), created in 2007 as a part of the seventh framework program. By allowing European researchers to compete with all other researchers in the EU area on the basis of excellence, ERC revolutionized the EU's approach to supporting scientific activity. With scientific excellence as the sole criterion for selection and sizeable pots of money to distribute (the first budget in the seventh framework program (2007–2013) was 7.51 billion euros, and it almost doubled for Horizon 2020, constituting 17 percent of the overall program budget),[21] ERC soon became a powerful tool for boosting European knowledge space and shaping science in Europe.

The shifting agendas and the broadening scope of the EC/EU activities have been paralleled by the gradual strengthening of the Community's legal status. After the still very restricted formulation of the Single European Act (1986), the Treaty of Maastricht (1992) included a new Article 130, which required that "the Community and the member states shall coordinate their research and technological development activities so as to ensure that national policies and Community policy are mutually consistent." Moreover, the Commission was granted the explicit right to take any "useful initiative" to promote such consistency.[22] The Amsterdam Treaty (1997) then abandoned the requirement for Council unanimity for adoption of the co-decision on the framework program. From now on, the program could be adopted by a qualified majority vote. The Union's competences in the domain were further widened by the Article 179 of Lisbon Treaty (2007), stating that:

1. The Union shall have the objective of strengthening its scientific and technological bases by achieving a European research area in which researchers, scientific knowledge and technology circulate freely, and encouraging it to become more competitive, including in its industry, while promoting all the research activities deemed necessary by virtue of other Chapters of the Treaties.

2. For this purpose the Union shall, throughout the Union, encourage undertakings, including small and medium-sized undertakings, research centres and universities in their research and technological development activities of high quality; it shall support their efforts to cooperate with one another, aiming, notably, at permitting researchers to cooperate freely across borders and at enabling undertakings to exploit the internal market potential to the full, in particular through the opening-up of national public contracts, the definition of common standards and the removal of legal and fiscal obstacles to that cooperation.

3. All Union activities under the Treaties in the area of research and technological development, including demonstration projects, shall be decided on and implemented in accordance with the provisions of this Title.[23]

Within ten years, the EC/EU research policy activity had firmly become woven into the legal fabric of the Union.

While the EC's/EU's legal base strengthened and its activities broadened, the Community/Union gradually moved into territories traditionally occupied not only by European nation-states but also by various intergovernmental organizations promoting research. In this book we have seen how already in the 1960s and 1970s, the EC challenged more established organizations such as the OECD and the Council of Europe. In the subsequent years, this trend continued. Indeed, as the EU is thus increasingly dominating the European research field, one can observe an incremental transformation of previously non-EC/EU projects and institutions into EU or quasi-EU projects and institutions, or alternatively, a launching of new EU operations practically identical to existing activities in other European organizations. A good example here is space research. The cooperation agreement signed between the EU and the European Space Agency in 2003 constituted a new framework for a comprehensive European space policy involving the coordination of the actions of the European Commission and ESA through a Joint Secretariat, a small team of EU administrators and ESA executives, as well as ministerial-level meetings in the Space Council. In 2007 there was a further development when twenty-nine European countries expressed their support for a European space policy, prepared jointly by the European Commission and the ESA's director-general. Moreover, the Lisbon Treaty gave the European Union an explicit competence in space, calling for the development of a European space program. The implementation of this program has further increased cooperation between the EU and the ESA. In the recent past, almost 20 percent of the funds managed by ESA have originated from the EU budget.[24]

The European Science Foundation, on the other hand, after almost nearing death in 2011,[25] has acknowledged the EU's growing role in promoting European science and wound down all its collaborative research instruments and networks. With a stated mission to serve the European Research Area for the sole benefit of science, it now focuses on running an expert division called Science Connect that provides services to the science community. This is a significant shift: since its creation in 1974, ESF has run over 2000 world-class science programs and networks, with support from eighty member organizations in thirty countries.[26] Through this work, it has made a solid contribution to the creation of the European research space. In the most recent years, its financial resources, however, have faced a dramatic downturn. Between 2010 and 2015, the ESF member contributions diminished by 15.4 million euros. These budget cuts, together with the wishes of its members, led ESF to radically reduce its activities and staff and to move policy-related work into newly formed Science Europe.[27] COST, another intergovernmental institution supporting European research set up in the early 1970s and analyzed in this book, has weathered the emergence of a comprehensive EU research policy somewhat better. It fills a specific niche in the European research funding landscape as an organization providing financial support for the creation of European-wide multi-stakeholder research networks, called COST Actions. Most of COST's funding, however, comes from the EU research framework program,[28] which makes it highly dependent on the political developments in the Union.

The various impacts of the emergent EU research policy on the wider European research space still deserve more thorough scholarly attention.[29] One interesting trend, however, is the emergence of stronger and more organized organizations and networks promoting the interests of researchers, research funders, and the higher education sector at the European level. These include EuroScience, a grassroots non-profit organization of researchers in Europe, created in 1997 to shape policies for science, technology, and innovation. With its 2600 individual members, EuroScience aims to be "a direct and democratic way to influence the construction of the Europe of Science and Technology, for the benefit of Europe's international position in science, and to enhance science's contribution to society, opportunities and mobility for both young and experienced researchers."[30] The European Association of Research and Technology Organizations (EARTO) was established two years later to promote research and technology organizations and represent their interest in Europe. Its current activities are geared toward influencing the EU policy.[31] In 2001, a merger between the Association of European Universities and the Confederation of European Union Rectors' Conferences led to the creation of the European University Association (EUA). EUA represents over 800 universities in forty-eight countries, and its activities are targeted at influencing the EU policies on higher education, research, and innovation.[32] A year later, another Belgium-based organization, the League of European Research Universities (LERU), was formed with the objective to "advance the understanding and knowledge of decision makers, policy makers and opinion leaders about the role and activities of research-intensive universities."[33] Both institutions seek to promote the interests of higher education and research institutions in EU policy-making. In 2011, twenty-three countries joined in the founding assembly of Science Europe, an organization representing research funders and other research organizations in the EU. Science Europe was built from two former advocacy groups, the ESF and EuroHORCS (an organization of the heads of the European research councils), aspiring to give its members a stronger and more united voice on European level policy.[34] The list could be continued, in particular if one includes more informal networks and alliances.

The rise of these various groups is largely a response to the gradual strengthening of the EU's research arm, which has made advocating for researchers' and research and higher education institutions' interests in Brussels a worthwhile endeavor. Undoubtedly, there is a new reality for research in Europe. The current EU research funding program is exceptional in its size, long duration (seven years), budgetary framework stability, and scope: it encompasses research as well as innovation; grants as well as loans, equity, and procurement; it combines a broad top-down focus on grand societal challenges with bottom-up "frontier" research; it is cross-border and cross-sectoral, encouraging inter-disciplinary collaboration, mobility, and coordination. And it is attracting increasing attention and interest within research communities both within and beyond Europe. The Horizon 2020 mid-term evaluation accounted a 65 percent increase in the annual number of applications compared to the program's predecessor, the seventh framework programme (FP7). With an 11.6 percent success rate, which was much lower than under FP7 (18.4 percent), the competition for Horizon 2020 grants had become fierce. While 92.8 percent of funding went to participants from the 28 EU member states, the program involved people from over 130 countries.[35] Moreover, EU-funded projects are also producing top-notch research. Between 2014 and 2016 alone, Horizon 2020, supporting approximately 340,000 researchers, produced more than 4000 peer-reviewed publications that were cited more than twice the world average. Two-thirds of them were also openly accessible to the public.[36] It is also noteworthy that within the European institutions, research has taken a prominent role. In 2016, a total of 1000 of the Commission's 23,000 officials worked for DG Research, making this DG the fourth largest division. If this number were to include staff members in the EU Joint Research Centre, research would occupy the first place. Research also continues to be the third largest item in the EU budget.[37]

Despite the prevailing uncertainties about the future of European integration and the different priorities of the member states, the trend of strengthening the role of the EU in supporting and shaping European research is likely to continue. In June 2018, the Commission published its proposal for the next EU funding program, Horizon Europe, which would cover the years of 2021–2027, involve a budget of 100 billion euros—an increase from Horizon 2020—and be "the most ambitious research and innovation funding programme ever."[38] There was little modesty in the Commission's statement that "'[b]eneficiaries' research capacities and scientific outputs would have significantly decreased had they received national funding instead. This decrease would have been especially large in terms of their ability to collaborate with industry and business, transfer of knowledge, the number of participations in scientific conferences and the knowledge in new areas."[39] The Covid-19 outbreak in 2020, effectively highlighting the importance of research in tackling global crisis is likely to give a further boost to EU research and innovation activities. The EU acted relatively fast by providing emergency funding from Horizon 2020 and processing proposals at a record speed.

One could note emerging tensions between European and national research spaces, as the national funding bodies and the EU instruments might increasingly compete for resources, applicants, and legitimacy.[40] The larger the EU research budget grows, and the more extensive the European level programs become, the more accentuated this competition will become. Already now the EU research funding programs include almost all research themes and forms of research promotion that exist in national research and technology policies.[41]

Linda Wedlin and Maria Nedeva explain the shift from "Science in Europe" to "European Science" by two important changes in policy rationales and framing. First, the understanding of "European added value" shifted from focusing on coordination to incorporating competition. Epitomized in the creation of the European Research Area, which signaled a move away from collaboration toward integration, this new framing precipitated the implementation of policy instruments that stretched beyond simple support for collaboration between researchers and encouraged genuine competition between them. Second, there was growing evidence and concern that Europe was not doing so well in science compared with its main commercial competitors. This new version of the technology gap debate argued for a European retard that was more fundamental than failure in making the link between scientific research and commercial applications. "By 2000," Wedlin and Nedeva write, "the long-standing assumption of this 'European paradox' morphed into a full-blown 'gap' argument, that is, that Europe was clearly lagging behind the USA and Japan both in terms of science and its application."[42] This concern led to a change of rationales and the target for policy intervention. It served as a justification for a powerful move to the realm of basic science and the creation of the European Research Council.

Tim Flink, through an analysis on the creation of ERC, supports this finding. A renewed concern of European global retard and a conviction of the need to strengthen the European knowledge base underpinned a political discussion that ushered the birth of ERC. Moreover, language played a critical role: with the concept "frontier research," the European Commission avoided directly violating the principle of subsidiarity, which had curtailed its activity in basic research—an area traditionally belonging to the nation states. "Frontier research," falling somewhere in between basic and applied research, aligned well with the EU's traditional mission of fostering European competitiveness and the geopolitical objective of the Lisbon Strategy to make Europe the world's leading commercial power, drawing its strength from knowledge. "Frontier research" projected an image of global competition, and it made a solid reference to Vannevar Bush's monumental report *Science, the Endless Frontier*[43] and the founding myth of the US National Science Foundation. "[I]n this policy process," concludes Flink, "the narrative structure only seemed to be successful if it followed the chimera of a geostrategic security understanding (isolation by the European Research Area) and a prosperity identity (market imperative of the EU)."[44]

Also, Terttu Luukkonen, while noting the critical role of stakeholder groups, such as the leading life sciences organizations, stresses the role of the gap debate in the formation of the ERC. "There was a concern," she writes, "about funding being too low for basic research and about quality of science and its institutions in Europe and, as in European research policy in general, USA provided a benchmark with which comparisons were made." Furthermore, the focus on frontier research, provided "a strong argument that the ERC is justified from the point of view of technological and economic competitiveness, not just of scientific competitiveness."[45] This is confirmed by just a cursory reading of the Commission's documents from that time period. The prominence of economic concerns and the gap narrative is striking, as well as the discourse that highlights the insufficiency of national initiatives. The Commission's proposal for ERA, for example, argues that in Europe, "the situation concerning research is worrying. Without concerted action to rectify this the current trend could lead to a loss of growth and competitiveness in an increasingly global economy. The leeway to be made up on the other technological powers in the world will grow still further. And Europe might not successfully achieve the transition to a knowledge-based economy."[46]

The discursive continuity is remarkable. Throughout this book we have seen how European Commission officials and other supporters of greater EU competences in the field of research used a specific language and political framing to advance their initiatives. This language presented scientific research as an engine for economic growth and material prosperity, which were widely accepted as favorable political goals by the European governing elites. Combined with arguments about European technological delay vis-à-vis its main commercial competitors, it presented an increasingly convincing case for new European level activity. By the early 1980s, these framings had become dominant truth claims in the European debates, and they started to essentially determine political agendas. At the same time, the scope and nature of the scientific enterprise changed, the pace of discovery and innovation accelerated, and as a result, not only did competition between research institutes, nations, and individual researchers accentuate, but science and technology became increasingly vital forces shaping societies and determining the geopolitical and economic success of countries.

It could be argued that one reason for the continuous popularity of the instrumental-economic conception of research policy has been its malleability and extraordinary ability to adapt to different political conditions. As amenable as it seemed to the Western European expansive welfare economies of the 1960s and 1970s, it has proved perfectly compatible with the intensifying drive for market liberalism since the early 1980s. This is not only because of the prevailing appeal of growth. Research policy has proved useful in facilitating and accelerating the shift to the world of open markets. In this study we have seen how research policy, especially at European level, was commonly pictured as a means to adapt to an exogenous and irreversible transformation that later became labeled as globalization.

The EU research policy came into being with an explicit objective of enhancing European economic performance in an international framework in which competition was accelerating. This goal remains as one of the guiding principles of the current EU research policy. When Research

Commissioner Máire Geoghegan-Quinn in November 2011 said: "fundamentally, support through Horizon 2020 for research, innovation and science is an economic policy,"[47] she reiterated the principle on which the Union's activity had been based since the very beginning. The economic framing remains sturdily in place. The Horizon 2020 mid-term evaluation report, for example, made a blatant calculation a GDP gain, concluding that "[e]very euro invested under Horizon 2020 brings back 6 to 8.5 euros."[48] The explanatory memorandum included in the Commission's proposal for establishing Horizon Europe included similar calculations. According to the memorandum, Horizon Europe was expected to generate positive effects on growth, with an increase of GDP on average by 0.08 percent to 0.19 percent over twenty-five years, "which means that each euro invested can potentially generate a return of up to 11 euro of GDP over the same period"—better ratio than in Horizon 2020. The memorandum went on to calculate that discontinuing the Union research and innovation program could result in a decline of competitiveness and growth up to 720 billion euros of GDP loss over twenty-five years.[49]

To some degree, the EU is tied to this language. Initially, the lack of a solid juridical basis for research policy forced the Commission to frame research as a tool for achieving prosperity, which has traditionally been one of the Union's core objectives. Although the gradual strengthening of the juridical basis of research policy during the last three decades has granted the Commission more room for maneuver, the EU's activity and ambitions are still limited by the principle of subsidiarity and the constant search for European added value. The EU is most likely to gain legitimacy by anchoring its initiatives and programs to the objectives of competitiveness and growth—a territory where EU activity is seldom questioned. But continuously liking research and growth and articulating science and innovation policy in economic terms, is more than a smart political strategy. It is a durable policy framing that was constructed in the early 1960s and that has been broadly embraced and absorbed by national decision-makers, European officials, and a number of other actors shaping research policy. It embodies a very specific understanding of the role of science and research in society and the objectives of public support for research and innovation communities.

Policy framings can and need to change if they become unsuited to responding to the society's most pressing needs and challenges. Johan Schot and W. Edward Steinmueller, for example, have argued for a new framing for innovation policy, "linked to contemporary social and environmental challenges such as the Sustainable Development Goals and calling for transformative change." This framing should take priority over the two existing framings, which emerged in the 1960s and 1980s, and which, while still relevant, "offer little guidance for managing the substantial negative consequences of the socio-technical system of modern economic growth to which they have contributed and of which they are a part."[50] The first framing, which dominated during the time period analyzed in this book and—as we have seen—still influences EU policy, emphasized the role of research and innovation for increasing economic growth, aimed at boosting the potential for science and technology for prosperity, and cultivated socio-technical systems supporting mass production and consumption. The second framing gained ground toward the late 1980s and more explicitly aimed to address challenges related to intensifying international competition and globalization. With a focus on developing and analyzing national systems of innovation, enhancing capacities for learning, and supporting connectivity between different societal sectors within the system, it added a layer of sophistication and complexity to the first framing. However, it didn't depart from the premises of increasing growth through maintained and improved competitiveness.[51]

The third framing suggested by Schot and Steinmueller draws upon the recognition that a deep and rapid transformation of socio-technical systems is needed in the backbone systems of modern societies, including energy, mobility, food, water, healthcare, and communication, and that research and innovation policy has a vital role to play in supporting this transformation. That requires a dramatic change in thinking, broadening of perspectives and alliances, and setting entirely new goals and objectives. According to the two authors, this new framing "focuses on innovation as a search process on the system level, guided by social and environmental objectives, informed by experience and the learning that accompanies that experience, and a willingness to revisit existing arrangements to de-routinize them in order to address societal challenges." Further, the innovation process in this framing "is likely to be effective in achieving these goals if it is inclusive, experimental and aimed at changing the direction of socio-technical systems in all its dimensions."[52] Essentially, this framing will propose building a new knowledge base, setting up mission-oriented policies with missions formulated in an open-ended way, encouraging new forms of engagement and networks between public, private, and third sector actors, and supporting intermediary actors to advocate competitive niches, as well as new visions and policies. It will mean a shift away from the old research and innovation policy proposition that social and economic goals can be achieved through economic growth if surpluses are redistributed and technocratic elites can regulate the externalities of growth. A more comprehensive and ambitious—if not radical—approach is needed to ensure the continued relevance of research and innovation policy in the contemporary world in rapid transition.[53]

Schot and Steinmueller are not alone with this proposition. In the last few years, there has been increasing scholarly interest in rethinking the premises of science and innovation policy and exploring ways of better aligning policy objectives with broader social and environmental challenges.[54] A new policy paradigm is emerging, "layered upon but not fully replacing earlier paradigms of science and technology and innovation systems policy." According to Gijs Diercks, Henrik Laren, and Fred Steward, this "'normative turn' that is currently taking place insists that innovation policy must not only optimize the innovation system to improve economic competitiveness and growth, but also include strategic directionality and guide processes of transformative change towards desired societal objectives." Further, this "emerging paradigm of transformative innovation policy is still a heavily contested discursive space" and "there is considerable uncertainty regarding the paths of this paradigm shift in the making."[55]

Looking at the most recent EU language on research and innovation policy, one can discern elements of this shift. While competitiveness and growth retain a solid place in the Framework Programs, with the increasing focus on strengthening the European knowledge base and the objective of tackling grand societal challenges through research, the political framing of the EU research policy has widened.[56] One of the earliest visible expressions of this ambition was the declaration, endorsed by over 300 researchers, policy-makers, and representatives from industry and research funding institutions during the 2009 "New World – New Solutions" conference in Lund, Sweden.[57] The declaration stated that "European research must focus on the Grand Challenges of our time moving beyond current rigid thematic approaches. This calls for a new deal among European institutions and Member States, in which European and national instruments are well aligned and cooperation builds on transparency and trust." The "Grand Challenges" listed in the declaration included global warming; tightening supplies of energy, water, and food; aging societies; public health; pandemics; and security. The Lund Declaration also states that the "European Knowledge Society must tackle the overarching challenge of turning Europe into an eco-efficient economy."[58] Since then, broader social and global concerns have found a firmer footing in the EU's research policy agenda.[59] Between 2014 and 2016, a total of 37 percent of Horizon 2020 funding went to reinforcing and extending the excellence of European science base and consolidating the ERA. The second biggest pot, 36 percent, went to stimulating a critical mass of research and innovation efforts to help address grand societal challenges. Promoting industrial leadership came only as third, with 22 percent of funding. The remaining 5 percent was dedicated to other priorities, such as widening participation, including society, Euratom, and the pilot for fast-tracking innovation.[60]

The three pillars of Horizon Europe—Excellent Science, Global Challenges and European Industrial Competitiveness, and Innovative Europe—seem to continue this trend, although the Commission notes that "[i]industrial leadership will be prominent in this [Global Challenges] pillar and throughout the program as a whole."[61] Indeed, the proposal for Horizon Europe was "framed by the premise that research and innovation (R&I) delivers on citizen's priorities, boosts the Union's productivity and competitiveness, and is crucial for sustaining our socio-economic model and values, and enabling solutions that address challenges in a more systematic way." In other words, "Horizon Europe will strengthen the Union's scientific and technological bases in order to help tackle major global challenges of our time and contribute to achieving the Sustainable Development Goals (SDGs). At the same time, the programme will boost the Union's competitiveness, including that of its industries."[62] These quotes reveal a striking feature in the Commission's blueprints for Horizon Europe: an effort to achieve the double objective of responding to global challenges and maintaining economic growth. Overall, however, the program seems to be defined as serving the European economy. In March 2019, Carlos Moedas, the Research Commissioner, noted that "ware now on track to launch the most ambitious ever European research and innovation programme in 2021, shaping the future for a strong, sustainable and competitive European economy and benefiting all regions in Europe."[63] Even the Lund Declaration underlines that "[m]eeting the Grand Challenges will be a prerequisite for continued economic growth."[64] A vital task for the Commission and the Union more broadly in the future will be resolving the tension between its traditional policy goals of pursuing growth and competitiveness and its newer mission of tackling grand challenges and promoting sustainability transitions. Can both objectives be successfully pursued in parallel, and how possible conflicts between growth and broader socio-environmental issues can be reconciliated, has not explicitly been addressed in any of the key documents or declarations—until very recently.

Ursula von der Leyen's Commission's proposal for a European Green New Deal contains explicit language, suggesting that the challenge of climate change can be addressed without compromising economic growth and competitiveness. It presents the European Green New Deal as "a new growth strategy that aims to transform the EU into a fair and prosperous society, with a modern, resource-efficient and competitive economy where there are no net emissions of greenhouse gases in 2050 and where economic growth is decoupled from resource use." Moreover, "careful attention will have to be paid when there are potential trade-offs between economic, environmental and social objectives." The language stressing the EU's "collective ability to transform its economy and society to put it on a more sustainable path," outlines an ambitious and transformative vision that departs from narrow sectorial propositions. In the European New Green Deal, science and innovation also has a solid place. The document calls for "new technologies, sustainable solutions and disruptive innovation" and notes that "conventional approaches will not be sufficient." The EU's new research and innovation agenda will emphasize experimentation, and working across sectors and disciplines, thereby taking the "systemic approach needed to achieve the aims of the Green Deal."[65] For now it seems that four of Horizon Europe's five research missions would be closely linked to the Green Deal objectives.[66] Also, how the Covid-19 pandemic transforms the European research landscape, remains to be seen.

In the last few decades, the EU has shown an important capacity of self-criticism and learning. Horizon Europe is being built on lessons learned in Horizon 2020, such as the need for increasing support for breakthrough innovation; adopting mission-orientation and encouraging citizen involvement; strengthening international cooperation; reinforcing openness; rationalizing the funding landscape; and encouraging participation. New initiatives in Horizon Europe, such as European Innovation Council, address these objectives.[67] Ursula von der Leyen's Commission has set a new level of ambition for achieving a pressing transformative change that transcends several sectors of European society. The rapid response to Covid-19 also shows proactivity and agility from an institution often seen as stiff and bureaucratic. These serve as demonstrations that the EU does not need to or even cannot be constrained by its past policy framings and discourses. And this is a real opportunity. In today's Europe, torn by nationalism and distrust in governing elites, preserving peace and producing prosperity no longer suffice to provide the legitimacy that carried the process of European integration this far. It is time for a more powerful and compelling mission, aiming at a profound socio-technological transformation that a Union-wide science and research effort is well positioned to shape and realize.

Despite its contested origins, research policy has become one of the EU's core instruments for advancing its policies, promoting European integration, and achieving change. It has won the support of researchers, both in Europe and beyond, gained prestige, and found a solid place on the EU political agenda. With its extensive programs and significant funding, it has transformed the European research landscape for good. However, to remain relevant, the EU research policy needs to be tailored to serve the most compelling needs of the European societies. If designed right, it can be a powerful force, enabling critical and truly transformative effort for which none of the individual member states would be capable alone.

NOTES

1. Maria Nedeva and Linda Wedlin, "From 'Science in Europe' to 'European Science,'" in Linda Wedlin and Maria Nedeva, eds. *Towards European Science. Dynamics and Policy of an Evolving European Research Space* (Cheltenham and Northampton: Edvard Elgar Publishing, 2015): 12–36.
2. European Commission: Science and Public Opinion (1977), available at: http://ec.europa.eu/public_opinion/archives/ebs/ebs_9_en.pdf (accessed December 25, 2019); European Commission, Qualitative Study on the Image of Science and the Research Policy of the European Union, October 2008, available at: https://ec.europa.eu/commfrontoffice/publicopinion/index.cfm/Survey/getSurveyDetail/yearFrom/1974/yearTo/2019/search/research%20policy/surveyKy/740 (accessed December 25, 2019).
3. A 2008 Commission survey found that among the European citizen, "the degree of knowledge on European research policy is extremely low." European Commission, Qualitative Study on the Image of Science and the Research Policy of the European Union, October 2008, available at: https://ec.europa.eu/commfrontoffice/publicopinion/index.cfm/Survey/getSurveyDetail/yearFrom/1974/yearTo/2019/search/research%20policy/surveyKy/740 (accessed December 25, 2019).

4. Katia Seidel, *The Process of Politics in Europe. The Rise of European Elites and Supranational Institutions* (London and New York: I.B Tauris Publishers, 2010): 154.

5. For a discussion on Europeanization from a historical perspective, see Martin Conway and Kiran Klaus Patel, eds. *Europeanization in the Twentieth Century: Historical Approaches* (New York: Palgrave, 2010).

6. John Peterson and Margaret Sharp, *Technology Policy in the European Union* (Basingstoke: Macmillan, 1998): 6, 70–78.

7. Peter Tindemans, "Post-war Research, Education and Innovation Policy-Making in Europe," in *European Science and Technology Policy, Towards Integration or Fragmentation?* Eds. Henri Delanghe, Ugur Muldur, and Luc Doete (Cheltenham and Northampton: Edward Elgar, 2009): 13–14.

8. Veera Mitzner, "Research Policy," *The European Commission 1986–2000. History and Memory of an Institution*, ed. Vincent Dujardin et al. (Luxembourg: European Commission, 2019): 321–334.

9. Linda Wedlin and Maria Nedeva, eds. *Towards European Science. Dynamics and Policy of an Evolving European Research Space* (Cheltenham and Northampton: Edvard Elgar Publishing, 2015): 23–25.

10. European Commission: Towards a European research area, Brussels, 18.1.2000, COM(2000) 6 final, available at: http://eurlex.europa.eu/LexUriServ/LexUriServ.do?uri=COM:2000:0006:FIN:EN:PDF (accessed December 7, 2019). For the development of ERA, see e.g. Thomas Banchoff, "Political Dynamics of the ERA," in *Changing Governance of Research and Technology Policy – The European Research Area*, eds. Jakob Edler, Stefan Kuhlmann and Maria Behrens (Cheltenham, UK and Northampton, MA: Edward Elgar, 2003).

11. Luis Sanz Memémdez and Susana Borràs, "Explaining Changes and Continuity in EU Technology Policy: The Politics of Ideas," in *The Dynamics of European Science and Technology Policies*, eds. Simon Dresner and Nigel Gilbert (Aldershot and Burlington: Ashgate, 2001): 41.

12. Terttu Luukkonen, "European Research Area: An Evolving Policy Agenda," in *Towards European Science. Dynamics and Policy of an Evolving European Research Space*, eds. Linda Wedlin and Maria Nedeva (Cheltenham/Northampton: Edvard Elgar Publishing, 2015): 37.

13. Banchoff, "Political Dynamics," 81–82.

14. European Commission Communication: A Reinforced European Research Area Partnership for Excellence and Growth, Brussels, July 17, 2012, available at: https://ec.europa.eu/research/science-society/document_library/pdf_06/era-communication-partnership-excellence-growth_en.pdf (accessed December 15, 2019).

15. Quoted in Luca Guzzetti, "The 'European Research Area' idea in the history of Community policy-making," in *European Science and Technology*

Policy. Towards Integration or Fragmentation? Eds. Henri Delanghe, Ugur Muldur, and Luc Doete (Cheltenham and Northampton: Edward Elgar, 2009): 73.

16. Luukkonen, "European Research Area," 39.
17. European Commission Communication: Europe 2020 Flagship Initiative Innovation Union, Brussels October 6, 2010, available at: http://ec.europa.eu/research/innovation-union/pdf/innovation-union-communication_en.pdf#view=fit&pagemode=none (accessed December 25, 2019).
18. European Commission Communication: A Reinforced European Research Area Partnership for Excellence and Growth, Brussels July 17, 2012, available at: https://ec.europa.eu/research/science-society/document_library/pdf_06/era-communication-partnership-excellence-growth_en.pdf (accessed December 25, 2019).
19. Inga Ulnicane, "Broadening Aims and Building Support in Science, Technology and Innovation Policy: The Case of the European Research Area," *Journal of Contemporary European Research*, 1 (2015): 31–49 (40).
20. Weldin and Nedeva, "Towards European Science," 25.
21. Tim Flink, *Die Entstehung des Europäischen Forschungsrates: Marktimperative – Geostrategie – Frontier Research* (Weiserswist: Velbrück Wissenschaft, 2016): 15.
22. Peterson and Sharp, *Technology Policy*, 116.
23. Consolidated version of the Treaty on the Functioning of the European Union—Part Three: Union Policies and Internal Actions, Title XIX: Research and technological development and space—Article 179 (ex Article 163 TEC), available at: https://eur-lex.europa.eu/legal-content/EN/TXT/?uri=CELEX%3A12016E179 (accessed December 25, 2019).
24. ESA website, https://www.esa.int/About_Us/Corporate_news/ESA_and_the_EU (accessed December 24, 2019).
25. Martin Hynes, "The European Science Foundation; Death or Mid-Life Crisis?" *Europhysics News* 46/1 (2015), 23–26.
26. The Science Connect Website, http://www.esf.org/esf/frequently-asked-questions/#content-1458 (accessed December 10, 2019).
27. Hynes, "The European Science Foundation," 24.
28. COST website: https://www.cost.eu/who-we-are/about-cost/ (accessed December 25, 2019).
29. Wedlin and Nedeva, "From 'Science in Europe,'" 30.
30. EuroScience website: https://www.euroscience.org/about/#about-us (accessed December 25, 2019).
31. EARTO website: https://www.earto.eu/about-earto/ (accessed December 25. 2019).
32. EUA website: https://eua.eu/about/who-we-are.html (accessed December 25, 2019).

33. LERU website: https://www.leru.org/mission (accessed December 25, 2019).
34. Daniel Clery and Gretchen Vogel, "European Research Heads Get a New Body," *Science*, October 21, 2011; "Science Europe lobby group hit by sudden exodus. Brussels-based advocacy group aimed to provide single voice for scientists in the EU—but is losing members," *Nature*, November 29, 2016.
35. European Commission, Key Findings from the Horizon 2020 Interim Evaluation, (Brussels: European Union, 2017), 3–4, 8, available at: https://ec.europa.eu/research/evaluations/pdf/brochure_interim_evaluation_horizon_2020_key_findings.pdf (accessed December 25, 2019).
36. Ibid., 10.
37. Tim Flink, "EU-Forschungspolitik—von der Industrieförderung zu einer pan-europäischen Wissenschaftspolitik?" In *Handbuch Wissenschaftspolitik*, eds. D. Simon et al. (Springer Reference Sozialwissenschaften, 2016), 80.
38. European Commission, "Proposal for a decision of the European Parliament and of the Council on establishing the specific programme implementing Horizon Europe—the Framework Programme for Research and Innovation," COM/2018/436 final—2018/0225 (COD), available at: https://eur-lex.europa.eu/legal-content/EN/TXT/?uri=COM%3A2018%3A436%3AFIN (accessed December 25, 2019); European Commission, "EU funding for Research and Innovation 2021–2027," Factsheet, June 7, 2018, available at: https://ec.europa.eu/commission/publications/research-and-innovation-including-horizon-europe-iter-and-euratom-legal-texts-and-factsheets_en (accessed December 25, 2019).
39. European Commission, Key Findings from the Horizon 2020 Interim Evaluation, (Brussels: European Union, 2017), 18, available at: https://ec.europa.eu/research/evaluations/pdf/brochure_interim_evaluation_horizon_2020_key_findings.pdf (accessed December 25, 2019).
40. Wedlin and Nedeva, "From 'Science in Europe,'" 31.
41. Flink, "EU-Forschungspolitik," 79.
42. Wedlin and Nedeva, "From 'Science in Europe,'" 23–25.
43. Vannevar Bush, *Science, the endless frontier*. A report to the President by Vannevar Bush, director of the Office of scientific research and development. (Washington D.C., United States Government Printing Office, 1945).
44. Flink, *Die Entstehung*, 20–21; 327–328. The author's translation from German.
45. Luukkonen, "European Research Area," 48–49, 51.
46. European Commission: Communication from the Commission to the Council, the European Parliament, the Economic and Social Committee, and the Committee of the Regions: Towards a European Research Area.

Brussels, January 1, 2000 COM(2000) 6 final, available at: https://eur-lex.europa.eu/legal-content/EN/TXT/PDF/?uri=CELEX:52000 DC0612&from=EN (accessed December 25, 2019).

47. Máire Geoghegan-Quinn, European Commissioner for Research, Innovation and Science: Remarks at press conference launching Horizon 2020 Press conference Brussels, November 30, 2011, Speech/11/033, available at: http://europa.eu/rapid/pressReleasesAction.do?reference=S PEECH/11/833&format=HTML&aged=0&language=EN&guiLangua ge=en (accessed December 25, 2019).

48. European Commission (Directorate-General for Research and Innovation), Key Findings from the Horizon2020 Interim Evaluation, (Brussels: European Union, 2017), 13, available at: https://ec.europa.eu/research/ evaluations/pdf/brochure_interim_evaluation_horizon_2020_key_find-ings.pdf (accessed December 25, 2019).

49. European Commission: Proposal for a regulation of the European Parliament and of the Council establishing Horizon Europe—the Framework Programme for Research and Innovation, laying down its rules for participation and dissemination, Brussels, June 7, 2018, COM(2018) 435 final, available at: https://eur-lex.europa.eu/legal-content/EN/TXT /?uri=COM%3A2018%3A436%3AFIN (accessed December 25, 2019).

50. Johan Schot and W. Edward Steinmueller, "Three Frames for Innovation Policy: R&D, Systems of Innovation and Transformative Change," *Research Policy* 47 (2018): 1554.

51. Ibid., 1554–1561.

52. Ibid., 1563.

53. Ibid., 1561–1565.

54. See, e.g. Egil Kallerud, "Goal Conflicts and Goal Alignment in Science, Technology and Innovation Policy Discourse," Working Paper (Oslo: Nordic Institute for Studies in Innovation, Research and Education, 2010); K. Matthias Weber and Harald Rohracher, "Legitimizing Research, Technology and Innovation Policies for Transformative Change. Combining Insights from Innovation Systems and Multi-Level Perspective in a Comprehensive 'Failures' Framework," *Research Policy*, 41 (2012): 1037–1047; Stefan Kuhlmann and Andre Rip, "Next Generation Innovation Policy and Grand Challenges," *Science and Public Policy*, 45 (2018): 448–454; Gijs Diercks, Henrik Larsen, and Fred Steward, "Transformative Innovation Policy: Addressing Variety in an Emerging Policy Paradigm," *Research Policy* 48 (2019): 880–894. In addition to scholars in STS and sustainability transitions studies, there have also been historians arguing for a change. For instance, Andrew Jamison wrote in 2014 that "as in the 1960s, there is a need for fundamentally rethinking the relations between science, technology, and society, in Europe as well as

internationally. In particular, there needs to be much more coordination between policies for science and technology and all the other policies that national governments, as well as local authorities and intergovernmental bodies pursue. In order to meet the challenge of climate change and sustainable development, science and technology need to be reconfigured so that the 'solutions' they provide can be relevant for the problems that humanity faces. And in order to provide appropriate solutions, scientists and engineers will need to be better educated about the problems that need to be solved." Andrew Jamison, "Science and Technology in Postwar Europe," in *The Oxford Handbook of Postwar European History*, ed. Dan Stone (Oxford: Oxford University Press, 2014): 647–648.

55. Diercks, Larsen, and Steward, "Transformative Innovation Policy," 881, 884.
56. Luukkonen, "European Research Area," 39; Ulnicane, "Broadening Aims," 31–49.
57. European Commission, Cordis: "Swedish Presidency: Research Must Focus on Grand Challenges," July 10, 2009, available at: https://cordis. europa.eu/article/id/31013-swedish-presidency-research-must-focus-on-grand-challenges (accessed December 15, 2019).
58. European Commission: *The Lund Declaration: Europe Must Focus on the Grand Challenges of our Time*, July 2009.
59. Luukkonen, "European Research Area," 44.
60. European Commission (Directorate-General for Research and Innovation), Key Findings from the Horizon 2020 Interim Evaluation, (Brussels: European Union, 2017), 3–4, available at: https://ec.europa.eu/ research/evaluations/pdf/brochure_interim_evaluation_horizon_2020_ key_findings.pdf (accessed December 25, 2019).
61. European Commission Factsheet, EU Budget for the Future. Research and Innovation, May 2, 2018, available at: https://ec.europa.eu/commission/sites/beta-political/files/budget-proposals-research-innovation-may2018_en.pdf (accessed December 15, 2019).
62. European Commission: Proposal for a regulation of the European Parliament and of the Council establishing Horizon Europe—the Framework Programme for Research and Innovation, laying down its rules for participation and dissemination, Brussels, June 7, 2018, COM(2018) 435 final, available at: https://eur-lex.europa.eu/legal-content/EN/TXT /?uri=COM%3A2018%3A436%3AFIN (accessed December 25, 2019).
63. European Commission, Presentation, Horizon Europe, The Next EU Research and Innovation Investment Programme (2021–2027), May 2019/Version 25, available at: https://ec.europa.eu/info/sites/info/ files/research_and_innovation/ec_rtd_he-presentation_062019_en.pdf (accessed December 15, 2019).

64. European Commission, *The Lund Declaration*, July 2009.
65. European Commission: Communication from the Commission to the European Parliament, the European Council, the European Economic and Social Committee, and the Committee of the Regions, The European Green Deal, Brussels, December 11, 2019, COM(2019) 640 final, available at: https://ec.europa.eu/info/sites/info/files/european-green-deal-communication_en.pdf (accessed December 25, 2019).
66. EU Council Meets to Debate Green Deal Grand Plan for Tackling Climate Change, *Science Business*, December 12, 2019, available at: https://sciencebusiness.net/news/eu-council-meets-debate-green-deal-grand-plan-tackling-climate-change (accessed December 25, 2019).
67. European Commission, Presentation, Horizon Europe, The Next EU Research and Innovation Investment Programme (2021–2027), May 2019/Version 25, available at: https://ec.europa.eu/info/sites/info/files/research_and_innovation/ec_rtd_he-presentation_062019_en.pdf (accessed December 15, 2019).

The manufacturer's authorised representative in the EU is Springer
Nature Customer Service Centre GmbH, Europaplatz 3, 69115 Heidelberg,
Germany. If you have any concerns regarding our products, please
contact ProductSafety@springernature.com

Printed and bound by CPI Group (UK) Ltd, Croydon, CR0 4YY

24/04/2026

02096315-0004